W9-AEP-420

THE MATH GENE

THE MATH GENE

How Mathematical Thinking Evolved and Why Numbers Are Like Gossip

KEITH DEVLIN

A Member of the Perseus Books Group

First published in Great Britain in 2000
by Weidenfeld & Nicolson

Published by Basic Books,
A member of the Perseus Books Group

A CIP catalog record for this book is available from the Library of Congress.

ISBN 0-465-01618-9

FIRST EDITION
00 01 02 03 / 10 9 8 7 6 5 4 3 2 1

CONTENTS

LIST OF ILLUSTRATIONS

ACKNOWLEDGMENTS

ALTHOUGH this book's thesis is consistent with all the evidence that I am aware of, I knew from the start it would be controversial. Some of the scientific territory I cover is hotly debated. Accordingly, I showed a few appropriate individuals my manuscript before sending it out for publication, so that I could test how well I had made my case. Derek Bickerton, Margaret Wertheim, Jeff Raskin, and J. D. Phillips all read it and offered useful suggestions, and I wish to express here my gratitude to them.

My debt to Bickerton is particularly significant. I first started thinking about the questions addressed here about twelve years ago and began to write a first draft about five years ago. As I started to assemble the various pieces of the puzzle, I realized that an important piece was missing. That missing piece was eventually supplied by Bickerton, through his book *Language and Human Behavior*. I had consulted it when it was first published, since I wanted to present Bickerton's account of the emergence of

Hawaiian creoles in my own book *Goodbye Descartes,* which I was writing at the time. In 1997, when I was starting to write this present volume, I read Bickerton's book more thoroughly. Once I had fully understood his account of how and why the human brain acquired language, everything fell into place. (Bickerton, in turn, says he was inspired in part by William Calvin.)

At the same time that Bickerton was unknowingly supplying the key piece of my puzzle, Stanislas Dehaene's excellent book *The Number Sense* came out. Although our two books address very different questions—Dehaene's focus is numerical ability, while my argument deals mainly with what is sometimes called "higher" mathematics (i.e., logical reasoning about formally defined patterns or abstract structures)—in one place there is considerable overlap. To develop my thesis, I had to look at the way the mind handles number—first, because the capacity to handle numbers is an ingredient of mathematical ability (and not just because arithmetic is part of mathematics), and second, because I wanted to show how *mathematical thought* differs from numerical computation.

The Number Sense included much of what I had to say on the way we handle numbers. Inspired by it, I extended my coverage of this topic a little by incorporating, in relevant places, some of Dehaene's examples. As a result, much of Chapters 2 and 3 now amounts to a summary of some parts of Dehaene's book. However, Dehaene says far more about the way the brain handles number than I do here—and provides more detail on the topics I do cover—and I would strongly urge any reader of my book to read *The Number Sense.*

When my completed manuscript was in the hands of my agents, Brian Butterworth's book *The Mathematical Brain* was published in the United Kingdom. (It's now available in the United States under the title *What Counts.*) Butterworth, like Dehaene, is a cognitive scientist, and, his title notwithstanding, his primary interest is also arithmetic, not (higher) mathematics. However, Butterworth's book overlaps with my account less than Dehaene's does—far less than his book and Dehaene's overlap with each other. I did add one or two examples from Butterworth, where they seemed to strengthen my case, but on the whole he and I follow different paths to different objectives.

Finally, my agents Diana Finch of the Ellen Levine Agency in the United States and Bill Hamilton of the A. M. Heath Agency in London were both enthusiastic proponents of the project from the start, and my editors William Frucht at Basic Books and Toby Mundy at Weidenfeld & Nicolson were both keen supporters. Bill Frucht, in particular, labored long and hard to help me craft the manuscript into a form that would be accessible and appealing to as wide an audience as possible.

Keith Devlin
Moraga, California
December 1999

PROLOGUE
The Wings of the Eagle

THE EAGLE HAS LANDED

THE VOICE from the spaceship *Eagle*, just above the surface of the moon, was clearly understandable despite the static. The other voice came from the Johnson Space Flight Center in Houston, Texas, and, not surprisingly, was much clearer. Both men spoke in short, efficient, factual statements, without outward emotion.

EAGLE: 35 degrees. 35 degrees. 750, coming down at 23. 700 feet, 21 down. 33 degrees. 600 feet, down at 19...540 feet...400...350 down at 4.... We're pegged on horizontal velocity. 300 feet, down 3½ . . . a minute. Got the shadow out there...altitude-velocity lights. 3½ down, 220 feet. 13 forward. 11 forward, coming down nicely...75 feet, things looking good.

HOUSTON: 60 seconds.

EAGLE: Lights on. Down 2½. Forward. Forward. Good. 40 feet, down 2½. Picking up some dust. 30 feet, 2½ down. Faint shadow. 4 forward. Drifting to the right a little.

HOUSTON: 30 seconds.

EAGLE: Drifting right. Contact light. Okay, engine stop.

HOUSTON: We copy you down, Eagle.

EAGLE: Houston, Tranquillity Base here. The Eagle has landed.

I was a twenty-two-year-old graduate student when Neil Armstrong and Buzz Aldrin landed on the moon. Today, thirty years on, I still get a deep thrill whenever I read the transcript of those last few minutes of the descent, as Armstrong guided the lunar landing craft—which the crew had christened the *Eagle*—to the spot where humans would first set foot on another planet.

As the culmination of a decade-long struggle to put a man on the moon (and bring him back alive), the actual landing did not represent a huge technological advance over the previous Apollo missions. To use Armstrong's own words, as he made the historic first step onto the moon's surface, the Apollo 11 moon landing was just "one small step for [a] man." But the symbolic importance of the event was inescapable. It was, as Armstrong went on to say, "One giant leap for Mankind." Although often touted as an achievement for science and engineering—which it was—I have always felt that the Apollo 11 mission was much more a triumph of the human spirit, and of two mental abilities that on the face of the earth are unique to humans: mathematics and language.

Getting to the moon was heavily dependent on mathematics, which underpins all the sciences and all of engineering. Every aspect of the mission was calculated(!) to the finest detail: how much fuel to carry for each stage of the flight to the moon; what path to take so that fuel would not be wasted in correcting the path; how much fuel would be required for the landing; how much for taking off again; how long each engine burn should last; how much oxygen was needed to keep the crew alive. The dialogue between *Eagle* and ground control during the final descent was almost entirely mathematical. They had little room for error: after Armstrong finally maneuvered the *Eagle* to a safe landing, there was just ten seconds' worth of fuel left.

So, it took mathematics to get to the moon. But where did language come in? Why do I say the moon landing was a triumph for language? Because the Apollo mission was a huge collaborative enterprise involving the coordinated efforts of thousands of people. Although only two men made those historic first steps on the moon, the project involved many thousands of individuals, spread all over North America—and elsewhere on the globe, if you include the people manning the tracking stations. Language provided the thread that tied the team together and allowed them to coordinate their actions to produce a single event.

Of course, it doesn't require a moon landing to remind us that language and mathematics are powerful tools. Both have led to countless other achievements that have transformed not only humankind but much of our planet.

One of my aims in this book is to convince you of just how remarkable and powerful—and uniquely human—language and mathematics are. Let me once again quote Neil Armstrong. When the lunar module broke free from the command ship that would remain in orbit above the moon during the course of the moon walk, Armstrong declared that "The *Eagle* has wings." The acquisition of language and mathematics gave humanity the wings to soar above our fellow creatures.

My other aim is to argue that these two faculties are not separate: both are made possible by the same feature of the human brain.

Along the way, I shall examine the questions of what exactly is mathematics, what exactly is language, and how they arose. I shall also consider a third, distinctly human faculty: our ability to formulate—and follow—complicated plans, worked out in advance, incorporating various alternatives to be followed, depending on how things turn out at the time. This faculty is closely linked with our abilities to use language and to do mathematics. And it too played a major role in the Apollo missions, every detail of which was worked out well in advance, with every imaginable contingency accounted for.

For example, the main flight plan called for the on-board computer to land the *Eagle* at a site determined months before the flight. But in the event, the lunar crew saw that the chosen site was uneven and littered with boulders. As they descended toward the moon's surface, Armstrong overrode the computers and landed the spacecraft manually. The possibility of a manual landing had been considered well in advance, and the Apollo crew had trained for it.

This last human ability—to imagine the future taking several different paths, and to make adaptable plans in response to our imaginings—is, in essence, the source of the other two (mathematics and language). Arguably, therefore, it is the most important of all.

THE MATH GENE?

Before we begin, I should clear up one thing: there is no "math gene" in the sense of a specific sequence of human DNA that confers mathematical ability. There are, of course, genes that affect our ability to do mathematics. But, in calling this book *The Math Gene*, I am simply adopting a common metaphor. Roughly speaking, by "the math gene" I mean "an innate facility for mathematical thought," just as authors sometimes use "the language gene" to refer to our innate facility to acquire and use language. Of course, both facilities are genetically determined (at least in part), as is almost everything else about us. But talk of a single "gene" for mathematics is purely metaphorical—as it almost always is when we read of "a gene for X."

My argument that you possess the math gene—i.e., that you have an innate facility for mathematics—is simply this: your genetic predisposition for language is precisely what you require to do mathematics. Now, the chances are that you have a great command of your native language but feel far less confident about your mathematical ability. Indeed, like millions of your fellow humans, you may even be "math phobic." Thus, in making my case, I will have to explain why many people seemingly cannot use a basic ability that I am claiming they possess. Part of the explanation is that most people do not really know what mathematics is. So I must also explain what mathematicians—people like myself—think of as "mathematics." It's not just numbers and arithmetic. Once you know what mathematics is really about, and once you see how our brains create language, you should find it far less surprising that thinking mathematically is just a specialized form of using our language facility.

·1·

A MIND FOR MATHEMATICS

I HATED MATHEMATICS when I was in elementary school. In 1971, at the age of twenty-four, I emerged from university a Ph.D. in mathematics and have been a professional mathematician ever since. The crucial change came when I was fifteen, when I discovered what mathematics was really about.

Although I hope to give you some idea of what I discovered and why I fell in love with it, that is not my main aim in this book. Rather, I want to solve a puzzle that has intrigued me for most of my adult life: how did our ancestors acquire a mind for mathematics?

By answering this question, we may begin to understand why so many people find mathematics impossibly hard.

NOT ENOUGH TIME

Why do I think that the acquisition of mathematical ability presents a puzzle? It's a question of time. The human species has had a recognizable concept of abstract numbers for at most 8,000 years. Formal, symbolic mathematics with equations, theorems, and proofs is little more than 2,500 years old. Calculus was developed in the seventeenth century; negative numbers came into widespread use in the eighteenth, and modern abstract algebra, where symbols like x, y, and z denote arbitrary entities, is a mere 150 years old.

Yet 8,000 years is nothing like enough time for the human brain to undergo major evolutionary developments. The 2,500-year history of formal mathematics is a mere eye-blink in evolutionary time. It has taken the human brain over 3,500,000 years to evolve to its present state. Einstein's brain differed little from that of Iron Age Man. Yet nothing we would call mathematics existed in the Iron Age.

Whatever features of our brain enable (some of) us to do mathematics must have been present long before we had any mathematics. Those crucial features, therefore, must have evolved to fulfill some other purpose. What purpose? (I'll phrase this more carefully later on.) My answer to this question has a startling consequence: everybody has the math gene.

The reason is simple to state: the features of the brain that enable us to do mathematics are the very same features that enable us to use language—to speak to others and understand what they say.

If I am right, then another question arises. Why are so many people seemingly unable to do math? If the basic facility to do mathematics is the same one that enables us to speak and understand speech, how come so few are able to use that facility for mathematics, given that every four-year-old is fluent with language? This second question requires us not only to look into the very heart of both language and mathematics, but also to ask ourselves what is the purpose of language—how did *it* evolve?

To develop my thesis, not only must I describe the nature of mathematical thought—mathematics is a woefully misunderstood subject—but I must also examine the way the brain works and how it evolved to its present state.

I shall also have to look deeply into the nature of language. I will have to explain the difference between language, which appears to be unique to humans, and a system of communication, which many species possess in varying degrees of complexity.

Much of my account will differ little from what you will find in most books on human evolution. But occasionally, the standard account does not fit the known facts. In particular, the most common reason given for the evolution of language is that it was driven by the need for ever greater communication—that communication was its original purpose, if you like. I find this explanation problematic. Far more likely, I will suggest, is that language first arose as a by-product of a growth in the brain's representational power. Only after language had arrived on the scene did its use in communication become a major selection factor.

The main activity that prepared the human brain for being able to do mathematics, I will suggest, was nothing to do with the physical world, as you might have expected; rather it was keeping track of interpersonal relationships in an increasingly complex society. Although you may find this suggestion surprising, it is not in conflict with other authorities on the issue, for the simple reason that no one else has yet attempted to explain how our capacity for mathematical thought evolved! (Numerical ability, yes; but not mathematical ability.) If my explanation is correct, we can begin to understand why so many people find mathematics so hard, and how we might modify mathematics instruction so that a greater proportion of people may learn it.

That then is the journey in front of us. How likely is it that I am correct?

As a mathematician, I feel confident when describing mathematics. I cannot bring the same confidence to the evolution of the human brain. But then, no one can. Piecing together a coherent account of the way our species evolved is a tricky business. The experts often disagree significantly about what happened, when, and why. No evolutionary argument can really escape the criticism that it is merely a "Just So Story." (The less loaded term is "rational reconstruction.") Of course, it is important (and I try) to be consistent with what is known for certain. But the known facts admit a wide range of possible interpretations.

IT'S NOT JUST NUMBERS

One consequence of our investigation is that we shall discover why people fall largely into two groups: those who find mathematics almost completely incomprehensible, and the much smaller group for whom it seems to come fairly easily—with almost no one in between those extremes. It turns out that those who can do mathematics have a secret.

Other questions we shall be able to answer are:

- Can you use language to help you be better at math? (Yes.)
- Do *mathematicians* think in language? (No.)
- What does it feel like to a mathematician to do mathematics?
- Do mathematicians have different brains? (No.)

Before we go any further, I should make it clear that this is a book about mathematics, not arithmetic. There's a big difference. Arithmetic is a part of mathematics, but most mathematics is not about arithmetic.

Since arithmetic is what schools teach first, and since many people stop learning mathematics before they have progressed to anything else, it's not surprising that the words "mathematics" and "arithmetic" are often taken to be synonymous. But in fact, the more advanced parts of mathematics have little to do with arithmetic or numerical computation, or even with numbers at all in the usual sense. Indeed, some of the best mathematicians are not very good with figures.

Still, much of what I shall say about mathematics has implications for arithmetic. In particular, as our story develops, we shall discover why many people have trouble with arithmetic.

For instance, we shall see why so many individuals have difficulty learning their multiplication tables, particularly the answers for 8×7, 9×6, and 9×8. (It is not too inaccurate to say that their problem stems from the brain's being too smart, not too dumb, and that the key to mastering the tables is to override the brain's natural intelligence.)

Other questions about arithmetic that we shall be able to answer are:

- Why do so many people dislike mathematics so much?
- Do Chinese and Japanese children have a built-in advantage over American and European children when it comes to learning mathematics? (Yes.)
- Do any animals have a sense of number? (Yes, many do, up to a point.)
- Can any animals do arithmetic? (Ditto.)
- What about newborn babies? (Yes. This fascinating answer required considerable ingenuity to obtain.)

Clearly, it's quite a journey we are about to embark on. What we discover along the way will not only tell us a lot about ourselves and about the natures of language and of mathematics, it also has significant implications for mathematics education.

Before we take the first step, however, I had better make good on my promise to tell you just what mathematics is. What does the word *mathematics* mean to a mathematician?

WHAT IS MATHEMATICS?

What is mathematics? If you ask this question of the first person you meet on the street, you will most likely hear that "mathematics is the study of number." If you insist that your respondent be more specific, you may elicit the suggestion that mathematics is "the *science* of numbers." But that is about as far as you will get, and it is not an adequate description of mathematics. It is out of date by 2,500 years! The answer to the question "What is mathematics?" has changed several times since then.

Until around 500 BC, mathematics was indeed about numbers. Ancient Egyptian, Babylonian, and Chinese mathematics consisted almost solely of arithmetic. It was largely utilitarian and very much of a "cookbook" variety. ("Do such and such to a number and you will get the answer.")

Between 500 BC and AD 300, mathematics expanded beyond the study of numbers. The mathematicians of ancient Greece were concerned more with geometry. Indeed, they regarded numbers in a geometric fashion, as measurements of length, and when they discovered that there were lengths to which their numbers did not correspond (called irrational lengths), their study of numbers largely came to a halt. For the Greeks, with their emphasis on geometry, mathematics was about numbers *and shape*.

Only with the Greeks did mathematics change from a collection of techniques for measuring, counting, and accounting into an academic discipline having both aesthetic and religious elements. At the start of the Greek period, Thales introduced the idea that the precisely stated assertions of mathematics could be logically proved by formal argument. For the Greeks, this approach culminated in the publication, around 350 BC, of Euclid's mammoth thirteen-volume text *Elements,* reputedly the second most widely circulated book of all time after the Bible.

After the Greeks, although mathematics advanced in several parts of the world—notably in Arabia and China—its nature did not change until the middle of the seventeenth century, when Sir Isaac Newton (in England) and Gottfried Wilhelm Leibniz (in Germany) independently invented the calculus. In essence, the calculus is the study of motion and change. Before calculus, mathematics had been largely restricted to the static issues of counting, measuring, and describing shape. The new techniques to handle motion and change enabled mathematicians to study the motion of the planets and of falling bodies on earth, the workings of machinery, the flow of liquids, the expansion of gases, physical forces such as magnetism and electricity, flight, the growth of plants and animals, the spread of epidemics, and the fluctuation of profits. Mathematics became the study of numbers, shape, *motion, change, and space.*

At first, calculus was mainly directed toward the study of physics, and many of the great seventeenth- and eighteenth-century mathematicians were also physicists. But from about 1750 onward there was increasing interest in the mathematical theory, not just its applications, as mathematicians sought to understand what lay behind the enormous power of calculus. By the end of the nineteenth century, mathematics had become the study of numbers, shape, motion, change, space, *and the mathematical tools that are used in this study.* This was the start of modern mathematics.

The growth of mathematical activity in the present century can best be described as an explosion in knowledge. In 1900, all the world's mathematical knowledge would have fit into about 1,000 books. Today it would take maybe 100,000 volumes to contain all known mathematics. Not only have existing branches, such as geometry and calculus, continued to grow, but many quite new branches have sprung up. At the turn of the century, mathematics consisted of about twelve subjects: arithmetic, geometry, calculus, and so on. Today, there are between sixty and seventy distinct categories. Some subjects, like algebra or topology, have split into subfields; others, such as complexity theory or dynamical systems theory, are completely new.

THE SCIENCE OF PATTERNS

Given such diversity, how does today's mathematician answer the question, "What is mathematics?" The most common answer is that mathematics is the *science of patterns.* This is fine once you understand exactly what mathematicians mean by "patterns" and how they go about examining them.

As this book progresses, we will examine several examples of different branches of mathematics. In the meantime, let me mention that the patterns studied by the mathematician can be either real or imagined, visual or mental, static or dynamic, qualitative or quantitative, utilitarian or recreational. They arise from the world around us, from the depths of space and time, and from the workings of the human mind. Different kinds of patterns give rise to different branches of mathematics. For example, number theory studies (and arithmetic uses) the patterns of number and counting; geometry studies the patterns of shape; calculus allows us to handle patterns of motion; logic studies patterns of reasoning; probability theory deals with patterns of chance; topology studies patterns of closeness and position.

Because these patterns are, for the most part, highly abstract, their description and study require an abstract notation. For instance, the symbolic notation of algebra is the most appropriate means of describing general behavioral properties of addition and multiplication. The commutative law for addition, for example, could be written in English:

When two numbers are added, their order is not important.

However, it is more economically written as:

$$m + n = n + m$$

The complexity and abstraction of most mathematical patterns make anything other than symbolic notation prohibitively cumbersome to use. And so the development of mathematics has involved a steady increase in the use of abstract notations.

The first use of algebraic notation in mathematics seems to have been by Diophantus, who lived in Alexandria around AD 250. In his treatise *Arithmetica*, generally regarded as the first "algebra textbook," Diophantus used special symbols to denote the unknown in an equation and to denote powers of the unknown, and he employed symbols for subtraction and for equality.

Modern mathematics books are awash with symbols, but mathematical notation no more *is* mathematics than musical notation is music. A page of sheet music represents a piece of music, but the notation and the music are not the same; the music itself happens when the notes on the page are sung or performed on a musical instrument. It is in its performance that the music comes alive; it exists not on the page but in our minds. The same is true for mathematics. When read by a competent performer (that is, someone trained in mathematics), the symbols on the printed page come alive—the mathematics lives and breathes like some abstract symphony in the mind of the reader.

Of course, this similarity can only be taken so far. Although only someone well trained in music can read a musical score and hear the music in her head, it requires no special training to enjoy a musical performance. But the only way to appreciate most mathematics is to learn how to "sight-read" the symbols. Although the structures and patterns of mathematics resonate in the human mind every bit as much as the structures and patterns of music, human beings lack the mathematical equivalent of ears. Mathematics can only be "seen" with the "eyes of the mind." It is as if we had no other way to appreciate the patterns and harmonies of music than sight-reading.

For many people, the highly abstract notation is a deterrent to their understanding of mathematics. (It has been said that every equation an author includes in a "popular" science book halves its sales.) But without algebraic symbols, large parts of mathematics simply would not exist. The issue is a deep one having to do with human cognitive abilities. The recognition of abstract concepts and the development of an appropriate language are two sides of the same coin.

The use of a symbol such as a letter, a word, or a picture to denote an abstract entity goes hand in hand with the recognition of that entity *as an entity*. To use the numeral "7" to denote the number 7 we must recognize the number 7 as an entity; to use the letter *m* to denote an arbitrary whole number we must have the *concept* of a whole number. The symbol allows us to think about and manipulate the concept.

We shall come back to this issue later. In the meantime, let me give you an overview of the journey on which we are about to embark.

WHAT DOES IT TAKE TO MAKE A MATHEMATICAL MIND?

A number of mental attributes contribute to our ability to do mathematics. We shall look in detail at each of these attributes (which are not all independent from each other) as our story unfolds. In particular, we shall ask ourselves when and how our ancestors acquired each of these abilities, and how they combined to give mathematical ability. We shall also ask whether the inability to do mathematics stems from a lack of one of these abilities or if the problem lies with bringing them together in the right way—or with some other cause. For now, let me say a little about the most significant ones.

Number sense: Along with several other species, humans have a sense of numerosity. We recognize the difference between one object, a collection of two objects, and a collection of three objects. We also recognize that a collection of three objects has *more* members than a collection of two. This sense is not something we learn; we are born with it.

Numerical ability: Number sense, the ability to distinguish and compare small numerosities, does not require a concept of numbers as abstract entities or an ability to count. Numbers and counting are learned (although there is some evidence that counting has an instinctive basis). With considerable effort, chimpanzees and apes may be taught to count up to ten or so. But as far as we know, only humans are able to continue the number sequence indefinitely and to count arbitrarily large collections.

Algorithmic ability: An algorithm is a specified sequence of steps that lead to a particular goal—the mathematician's equivalent of a recipe for baking a cake. Doing arithmetic requires an ability to learn various sequences of operations on numbers. Other parts of mathematics require one to apply algorithms to other kinds of entities. For example, solving a quadratic equation involves following an algorithm of algebraic manipulations.

These three attributes provide most of the ingredients for being able to do arithmetic. Individuals who are good at arithmetic, however, often use additional attributes as well. For example, my childhood difficulty in learning my multiplication table—I was one of the last in the class to do it—became easier when I realized that I only needed to learn half the entries. If I knew that $7 \times 9 = 63$, then I could find the answer to 9×7 by applying the logical rule that order is not important in multiplying: 9×7 is the same as 7×9. To this day, I work out 9×7 by first turning it around to give 7×9 and then calling on my memory to retrieve the fact that $7 \times 9 = 63$.

The remaining attributes all contribute, to a greater or lesser extent, to mathematical (as opposed to arithmetic) ability.

The ability to handle abstraction: To my mind, a limitation in coping with abstraction presents the greatest barrier to doing mathematics. And yet, as I shall show, the human brain acquired this ability when it acquired language, which everyone has. Thus, the reason most people have trouble with mathematics is not that they don't have the ability but that they cannot apply it to *mathematical abstractions*. It will be an interesting challenge to explain how this situation evolved.

A sense of cause and effect: Like many other species, humans seem to acquire this sense at a very early age. Its survival advantage is obvious.

The ability to construct and follow a causal chain of facts or events: The ability to construct and follow fairly long causal chains appears to be unique to humans beyond the first few years of life. As I shall explain, our ancestors acquired this ability when they acquired language. The mathematician's proof (of a theorem) is a highly abstract version of a causal chain of facts.

Logical reasoning ability: This is the ability to construct and follow a step-by-step logical argument. It is closely related to the ability above, and is fundamental to mathematics.

Relational reasoning ability: Much of mathematics is about the relationship between (abstract) objects. I shall argue that reasoning about mathematical relationships between mathematical objects is no different from reasoning about physical relationships between physical objects or about human relationships between people. Since most of us engage in such reasoning every day, this once again raises the question why so many people find it hard to reason about mathematical objects.

Spatial reasoning ability: The ability to reason about space is crucial to many species' survival. This ability, which forms the basis for geometry, may also be used to reason about domains that are not, on the face of it, spatial. Indeed, many of the major discoveries in advanced mathematics stem from mathematicians finding novel ways to view problems in a spatial fashion. (The proof of Fermat's last theorem in 1994 came about in that way.)

Those, then, are the mental abilities that combine to give us the ability to do mathematics. Our quest for the origins of mathematical ability reduces in large part to a search for the origins of each of the abilities we have just considered. The framework for that search is human evolution. Each of the abilities listed above carries a cost in terms of the brain's energy consumption. (Some carry other costs as well.) Thus, it must have provided a survival advantage that outweighed the cost. In some cases—for example, spatial reasoning or a sense of cause and effect—that advantage is obvious. Other cases require us to dig deeper.

THE PUZZLE OF BRAIN SIZE

One aspect of the human brain that I did not list is size. Modern imaging techniques have shown that large parts of the brain are active when a person is doing mathematics. Thus, a large brain seems to be a crucial prerequisite for mathematical ability. The brain's structure—the number and nature of the connections between individual neurons—is also believed to play an important role.

How, then, did *Homo sapiens* come to have such a large and highly structured brain? The cost of such a large brain is enormous—the brain makes up less than 2 percent of the body's mass, yet uses about 20 percent of its energy—so its survival advantage must have been enormous as well. In which case, how is it that through the entire history of life on earth, only one species has developed a brain of anything like the size (relative to body weight) of the human brain?

The human brain is nine times larger than is normal for a mammal of our body size, and thirty times larger than the brain of a dinosaur of the same body size. Its actual size varies between 1,000 and 2,000 cubic centimeters, with the vast majority between 1,400 and 1,500 cubic centimeters. Within this range there is no obvious correlation between size and intelligence. Some very intelligent people have brains of around 1,000 cubic centimeters, and others who show no signs of what we would call high intelligence possess 2,000-cubic-centimeter giants. In fact the Neanderthals, those supposedly dim-witted hominids who died out about 35,000 years ago, had slightly larger brains than ours, with most falling in the range of 1,500 to 1,750 cubic centimeters.

In terms of relative brain size, our closest rivals are the dolphins and porpoises, and after them the non-human primates—apes, chimpanzees, and monkeys. But all lag well behind us in the brain-to-body-weight ratio.

Because we've inherited such a brain, we can provide a long list of things we can do with it: science (including mathematics), technology, art, and culture are all products of our individual and collective intelligence. But these are very recent developments. The hominid brain reached its present size about half a million years ago—long before science, art, or culture could have influenced its evolution. How, and for what purpose, did such a large and powerful brain evolve in the first place? And what enabled it to turn its power to mathematical thought some half a million years later?

This is the journey before us. Although our main focus is not our arithmetical ability, it is there that I shall begin. For just as numbers and arithmetic provide—for most of us—the gateway to the rest of mathematics, so too they provide a convenient place to start on our quest.

·2·

IN THE BEGINNING IS NUMBER

THE GREAT nineteenth-century German mathematician Leopold Kronecker once wrote, "God made the integers, all else is the work of man." (The integers are the positive and negative whole numbers.) His point was that, starting with the integers, it was possible to develop all of mathematics. Given that many contemporary branches of mathematics have little to do with numbers, his observation can be misleading when applied today. Nevertheless, the integers do play a fundamental role in mathematics. And, of course, they represent most people's first introduction to mathematics.

As I indicated in the previous chapter, our ability to handle numbers—to count collections and to do arithmetic—rests largely on three mental capacities: number sense, numerical ability, and algorithmic ability. How common are these abilities? To what degree do other species have the same or similar abilities? How and when did our ancestors acquire them, and what survival benefits did they confer?

In this chapter, we'll take a look at the number sense. In the following chapter, we'll move on to examine arithmetical ability.

Number sense holds a number of surprises. One is that, no matter how mathematically inept we think we are, every one of us has a built-in number sense and a rudimentary arithmetical ability. Another surprise is that babies exhibit these basic abilities when they're just a few days old. Still another surprise is that many animals, from the pigeon to the chimpanzee, possess similar number sense and arithmetical ability.

Unlike our ability to perform mathematical reasoning, our basic number sense seems quite independent of language. Thus the discussion of number in this chapter and the next is not a part of my overall argument about language and mathematics, which will come later. Yet there is a connection. As I shall describe in the pages that follow, we use our facility with language to extend our innate number sense and make numbers perform useful work for us.

THE NUMBER SENSE

So you think you don't have a head for figures? Okay, answer the following questions as fast as you can:

$$1 - 1 = ?$$

$$4 - 1 = ?$$

$$8 - 7 = ?$$

$$15 - 12 = ?$$

As soon as you have done that, pick a number between 12 and 5; any one, the first one that comes into your head.

Done that?

You picked the number 7, didn't you? How did I know? Because I knew you would follow your innate number sense. (If, despite my instructions, you did not pick *the first number that came into your head*, you might not have chosen 7; but the chances are that, even then, 7 was your choice.) Why 7?

Here is the explanation offered by the cognitive psychologists. The first four questions were all subtractions. Although they were all very easy, answering them got your mind into "subtraction mode." Then, when presented with the numbers 12 and 5, you subconsciously computed—or at least estimated—the distance between them, namely $12 - 5 = 7$, making the number 7 salient in your mind. (If you are one of the rare people who, despite following my instructions, did not give 7 as your answer, the chances are very high that you chose a number close to 7—namely, 6 or 8.)

Notice that 7 is not roughly halfway between 5 and 12; if that were the principle subconsciously guiding your choice, you would have picked 8 or 9. The number 7 lies toward one end of the given range. Why did you not pick 10, which is placed at the other end of the range? After all, 10 is much more common in daily life than 7, given that 10 is the base for our method of arithmetic.

When you think about it, there are several rational choices. But when you are first presented with the challenge, 7 is the number that spontaneously pops into your mind, the distance between 5 and 12.

Quick, what is 8 times 7? It's 54, right?

Or is it 64? Or 56, perhaps? If you are like most people, each of these answers seems "reasonable." Why is it that, despite hours of drill in elementary school, most of us have so much trouble with our multiplication table? Especially when each of the numbers being multiplied is 6, 7, 8, or 9?

Okay, one more. Here's a number comparison test. For each of the following pairs of numbers, say which is greater:

1 and 50

5 and 4

25 and 24

You may not have been aware of it, but if someone had been timing your response (say by asking you to press one of two buttons depending on which of the pair was the greater), you would discover that it took you longer to answer the second question than the first, and longer still to answer the third. Why?

You might explain that you answered the first more quickly because "the two numbers are so far apart, it's obvious that 50 is bigger than 1." But what's that got to do with it? I didn't ask you how far apart they were. I asked which was the greater.

How about the second and the third examples? Arithmetically, these are the same: each is a pair of successive numbers, with the greater written first. Indeed, if you simply ignore the first digits in the third example (which you may, since it's the same digit in both numbers), the two examples are identical. And yet it has been shown on many occasions that everyone takes measurably longer to decide between 25 and 24 than between 5 and 4. A difference of 1 is somehow more easily recognized for pairs of small numbers than for larger ones.

The fact is, you have an innate sense of number. You have had it since you were a few days old, and quite possibly you were born with it. And we humans share this number sense with chimpanzees, rats, lions, and pigeons.

I am not saying you are "good at sums" or that you "know your multiplication table." But regardless of your prowess in the mathematics class, your mind does have a sense for number. Numbers—at least small ones—have *meaning* for you, just as do words and music, and that meaning is not something you had to work at to learn. You were either born with it or born with a natural ability to acquire it effortlessly and unavoidably at a very early age.

The term "number sense" was introduced by Tobias Dantzig in his 1954 book *Number: The Language of Science*. He wrote:

> Man, even in the lower stages of development, possesses a faculty which, for want of a better name, I shall call *number sense*. This faculty permits him to recognize that something has changed in a small collection when,

without his direct knowledge, an object has been removed or added to a collection.

Stanislas Dehaene took the term "number sense" as the title of his recent book. It's a good book. It contains most of what you will find in this chapter, and a great deal more. I recommend that you read it. It does, however, have one flaw: its subtitle, "How the Mind Creates Mathematics." Not only does the kind of basic number sense that Dehaene focuses on have little to do with mathematics, but at no point does he even begin to tell us how the mind *creates* mathematics. Indeed, he barely scratches the surface of how the mind performs arithmetic, which is just one branch of mathematics, and an atypical one at that. This book, in contrast, is very definitely about all of mathematics. My intention is to understand how the human mind acquired the ability to perform mathematical reasoning.

Since mathematical thinking seems to be unique to humans, we may gain some insight into the key factors that led to it by comparing our mathematical ability with that of other species. In particular, to demonstrate that mathematical ability is nothing other than linguistic ability used in a slightly different way, it should be helpful to see what mathematical abilities are possessed by creatures that do not have language.

There is in fact a fairly extensive literature on laboratory studies of number sense in animals. (Dehaene lists some good sources in his book.) One of the first to realize that animals have a sense of number was the German psychologist Otto Koehler. During the 1940s and 1950s, Koehler suggested that two important prerequisites for arithmetic were the ability to compare the sizes of two collections presented simultaneously and the ability to remember numbers of objects presented successively in time. Both of these abilities form part of what I am here calling number sense. Koehler showed that birds have both abilities.

In one case, a raven called Jakob was repeatedly presented with two boxes, one of which contained food. The lids of the boxes had different numbers of spots arranged randomly. A card placed alongside the two boxes bore the same number of spots—although arranged differently—as the lid of the box with the food. Through many repetitions, the raven learned that to obtain the food, it had to open the box whose lid bore the same number of spots as the card. In this way, it was eventually able to distinguish two, three, four, five, and six spots.

In another experiment, Koehler trained jackdaws to open the lids of a row of boxes to obtain food until they had taken a given number of pieces, say four or five. Each box contained one, two, or no pieces of food, distributed randomly on each repetition, so there was no possibility of the birds basing their actions on a geographic feature such as the length of the row of boxes they opened. Rather, they had to keep an inner tally of how many pieces they had taken; in our terms, they had to count.

Another illustration of birds' numerical abilities comes from Irene Pepperberg, who trained her African Gray parrot Alex to say how many objects it saw on a tray, a task that required that the bird not only distinguish numerosities but also associate an appropriate vocal response with each number.

Many birds also exhibit a sense of numerosity in the number of times they repeat a particular note in their song. Members of the same species born and reared in different regions acquire a local "dialect," with the number of repetitions of a particular note varying from one location to another. Thus, although many aspects of a bird's characteristic song may be genetically determined, the number of repetitions of a particular note seems to be acquired by a young bird imitating the older birds around it, most likely its parents. For example, a canary raised in one area may repeat a particular note six times, whereas one raised elsewhere will repeat the same note seven times. Since the number of repetitions is constant for each bird, this means that the bird can "recognize" the number of repetitions in its song.

One obvious survival advantage to being able to compare numbers of objects in collections is that it helps a group of animals to know whether to defend their territory against an attack or to retreat. If the defenders outnumber the attackers, it might make sense to stay and fight; if there are more attackers, the wisest strategy might be to make a bolt for it. This suggestion was put to the test a few years ago by Karen McComb and her colleagues. They played tape recordings of roaring lions to small groups of female lions in the Serengeti National Park in Tanzania. When the number of different roars exceeded the number of lions in the group, the females retreated; but when there were more females, they stood their ground and prepared to attack the intruders. They seemed able to compare number across two different senses: the number of roars they *heard* versus the number of lionesses they *observed*, a task that seems to require a fairly abstract number sense.

THE HORSE THAT DIDN'T KNOW 2 + 2 = 4 AND THE RAT THAT DID

Claims that various animals have been shown to possess a number sense are sometimes dismissed by the experts. Much of the blame for this reaction can be laid at the feet of a horse living in Germany at the start of the twentieth century. After more than ten years' effort, a certain Wilhelm von Osten claimed to have taught arithmetic to his horse, Hans. Both horse and master became celebrities, and the German newspapers carried stories about "Clever Hans."

A typical demonstration would see von Osten and his horse surrounded by an eager audience. "Ask him what is three plus five," someone would call out. Von Osten would write the sum on chalkboard and show it to the horse, who would then carefully tap his hoof on the ground exactly eight times. Other times, von Osten would show Hans two piles of objects, say four in one pile and five in the other. Hans would tap his hoof nine times.

Even more impressive, Hans could apparently add fractions. If von

Osten wrote the two fractions $1/2$ and $1/3$ on the board, Hans would tap his hoof five times, then pause, then tap six times, to give the correct answer $5/6$.

Of course, there were suspicions of a trick. In 1904 a committee of experts gathered together to investigate the matter, among them the eminent German psychologist Carl Stumpf. After carefully observing a performance, the committee concluded that it was genuine—Hans really could do arithmetic.

One person, however, was not convinced by the committee's findings. Stumpf's student Oskar Pfungst insisted on further testing. This time, Pfungst wrote the questions onto the board himself, and he did so in such a way that von Osten could not see what was written. This enabled him to do something Stumpf had not. On some occasions, Pfungst wrote down the question that had been given to him. Other times, he changed it. Whenever Pfungst wrote down the question as given to him, Hans got it right. But when he changed the question, Hans gave the wrong answer —in fact, he answered the question von Osten *thought* had been given to the horse.

The conclusion was inescapable: von Osten had been doing the arithmetic. Through some subtle cue, perhaps a raised eyebrow or a slight shrug, he had been instructing Hans when to stop tapping his hoof. As Pfungst acknowledged, von Osten could well have been oblivious to this. Having worked so hard to train him, von Osten very much wanted his four-legged protégé to succeed. Doubtless, he became very tense as Hans's tapping got to the crucial number, and presumably Hans was able to detect some external manifestation of that tension. Thus, while Pfungst's investigation showed that Hans's performance did not require unusual arithmetical powers, it did show that humans could communicate with horses by means of the subtlest actions.

The case of Clever Hans showed the importance of proper design for any psychological experiment, to eliminate any possibility of subtle communication of clues. Unfortunately, the affair made subsequent claims of arithmetical abilities in animals extremely difficult to get taken seriously.

And yet nothing Pfungst did showed that animals could not have number sense. He simply showed that, *in Hans's case,* it was von Osten who had performed the calculations, not the horse.

In fact, a number of carefully conducted studies have shown that some animals can indeed perform some kind of arithmetic. One convincing series of demonstrations was carried out in the 1950s and 1960s by the American animal psychologists Francis Mechner and Laurence Guevrekian.

The idea was to deprive a rat of food for a short period and then put it into a closed box with two levers, A and B. Lever B was connected to a mechanism that delivered a small amount of food. But to activate lever B, the rat first had to depress lever A a fixed number (n) of times. If the rat depressed lever A fewer than n times and then pressed lever B, it received a mild electric shock and no food. Thus, to eat, the rat had to learn to press lever A n times and then press lever B.

At first, by trial and error, the rats discovered that, in order to get food, they had to press lever A a certain number of times and then press B. With repeated trials, they gradually learned to estimate the number of times they had to press lever A. If the apparatus was set up so that four presses of lever A were required to activate lever B, then, over time, the rats learned to press lever A about four times before pressing lever B.

The rats never learned to press lever A *exactly* four times on every occasion. In fact, they tended to overestimate, pressing it four, five, even six times. Given that they received an unpleasant shock if they pressed lever A fewer than four times, this "play safe" strategy makes sense. In any event, it did seem that the rats were able to estimate four presses. Likewise, in an apparatus set up so that lever A had to be pressed eight times, they learned to press it about eight times. In fact, they could learn to press lever A as many as sixteen times.

To avoid any possibility that the rats were judging time rather than the number of presses, the experimenters subsequently varied the degree of food deprivation. The more hungry the rats were, the more rapidly they pressed the lever. Nevertheless, rats trained to press level A four times continued to do so, and there was a similar effect with the rats trained to another number. Time was not the factor; they were estimating the number.

Notice that I have not said that the rats counted. What the experiment showed is that, through training, rats are able to adjust their behavior to press a lever *about* a certain number of times. They may have been counting, albeit badly. But there is no evidence for this. I think it's far more probable that they were simply *judging* or *estimating* the number of presses, and moreover doing so as well as we ourselves could if we did not count.

Other experiments performed on rats—some of which are described by Dehaene—point to the same conclusion: rats have a sense of number.

What evolutionary advantage led to selection for a number sense in rats? One possibility is the need to remember navigational information, such as its hole was the fourth one along after the third tree. It's also useful in keeping track of other animals in the vicinity, be they friends or predators.

As it happens, an experiment was performed that could be said to have addressed this very issue, but as things turned out, the result was not at all what was expected.

The experimenters presented their subjects with a corridor of doors, each of which led to food, and all but one of which were locked. For example, in a row of ten doors, only door number 7 might be unlocked. The experimenters wanted to see if the rats would learn to ignore the first six doors and go straight to door 7. The experiment seemed to be a huge success. After a number of initial learning cycles in which they tried one door after another, the rats seemed to figure out what was going on. Soon, each rat would race along the corridor at top speed until it reached door 7, and then push open the door to get the food.

When the experimenters watched a videotape of the rats in slow motion, however, they discovered that the rats were not counting at all.

As each rat dashed along the corridor, it gave each door a light push with its hind leg until it encountered one that yielded. Whereupon it screeched to a halt and shot into the room. This did not show that rats cannot count, nor that they could not have used a counting strategy. But that is not what they did. The lesson the experimenters learned was to be careful in ascribing an explanation to what has been observed. Things may not always be as they first appear.

WHAT ABOUT THE CHIMPANZEES?

So much for rats. What about chimpanzees? Given their similarity to humans, we might expect them to exhibit the best-developed number sense. Do chimpanzees in fact have any arithmetical ability? Guy Woodruff and David Premack of the University of Pennsylvania set out to investigate this in the late 1970s and early 1980s.

Woodruff and Premack started out by aiming high. In their first experiment, the two investigators showed that chimpanzees can understand fractions. For instance, they showed the chimp a glass half-filled with a colored liquid and then got the animal to choose between two further glasses, one half-filled, the other three-quarters full. The subjects had no difficulty mastering this task. But was the chimp basing its choice on the volume of the water in the glass or on the fraction by which it was full? The answer was obtained by making the task more abstract. This time, after the chimp was shown a half-full glass of liquid, for example, it would be presented with half an apple adjacent to three-quarters of an apple. The chimp consistently picked the half-apple over the three-quarter apple. The same thing happened when the chimp was shown half a pie against one-quarter of a pie. In fact, when presented with any choice between one-quarter, one-half, and three-quarters, the chimp was able to spot the correct fraction. It knew, for instance, that one-quarter of a glass of milk is the same fraction of a whole glass that one-quarter of a pie is of a complete pie.

Going a step further, Woodruff and Premack first presented two initial stimuli, say a quarter of an apple together with half a glass of liquid, and then asked the chimp to choose between a full pie and three-quarters of a pie. The chimp tended to choose the three-quarters pie—not always, but too often for its choices to be random. Apparently the chimp could perform the computation $\frac{1}{4} + \frac{1}{2} = \frac{3}{4}$, at least in an approximate, intuitive fashion.

Many other experiments show that chimps possess an ability in basic arithmetic. For example, a chimp is presented with two alternative choices of a treat. On one tray are placed two piles of chocolates, one with three chocolates, the other with four. The alternative tray has one pile of five chocolates together with one additional chocolate by itself. The chimp can only choose one tray. Which one does it choose? If it bases its choice on the largest pile it sees, it should pick the tray with the pile of five chocolates. But if it can add up the total number of chocolates on each tray, it will realize that the first tray has seven chocolates whereas the other has only six. Most of the time, without any special training, the chimp will select the tray with seven chocolates. The chimp can estimate that $3 + 4 = 7$ and $5 + 1 = 6$, and moreover can tell that 6 is less than 7.

The numerical approximating ability shown by rats and chimpanzees resembles the innate numerical estimating ability in humans. But humans can do more. We can count precisely and perform exact arithmetic. One of the keys to these skills is that we can use symbols to denote numbers. Arithmetic can then be performed in an essentially linguistic fashion, by manipulating symbols according to precise rules. Can chimpanzees be taught symbolic notation?

The answer is yes—up to a point. In the 1980s, Tetsuro Matsuzawa, a Japanese researcher, taught a chimpanzee named Ali how to use correctly the Arabic numerals 1 to 9. Ali was able to use these numerals to give the number of objects in a collection with up to 95 percent accuracy. Based on his response times, it appears that Ali could recognize at a glance the number of objects when they were three or fewer, but resorted to counting for larger collections. Ali could also order the numerals according to their magnitude.

A number of subsequent investigations have produced similar results. In one of the most impressive, Sarah Boyson provides her chimp Sheba with a collection of cards, on each of which is printed a single digit between 1 and 9. Sheba can correctly match each of the printed digits with a collection of between one and nine objects presented to her. Sheba can also do simple addition using symbols. For instance, if Boyson holds up the numerals 2 and 3, Sheba can successfully pick out the card bearing the numeral 5. More recently, in 1998, Elizabeth Brannon and Herbert Terrace have demonstrated a similar ability of rhesus monkeys to learn to distinguish the numbers 1 to 9 and to associate symbols with those numbers.

I should stress that it took many years of slow and painstaking training to achieve the kind of performance exhibited by Sheba and the various other chimpanzees, monkeys, dolphins, and what-have-you in such experiments. Teaching animals the link between the abstract symbols and collections of objects is a long and arduous process. Performance is never perfect and is limited to very small collections. Young children, by contrast, take just a few months to catch on to numbers. And once they do, they do so in a big—and accurate—way. When it comes to numbers, humans are very different from all other animals, and the difference shows up at a very early age.

THE RISE AND FALL OF PIAGET

Much of our current popular wisdom about small children's mental abilities originates in the work of the cognitive psychologist Jean Piaget fifty years ago. Piaget's influence can be found not only in many of our current beliefs about the way children learn, but also in our educational systems. Unfortunately, as often happens with ground-breaking research, subsequent investigations have shown that many of Piaget's conclusions were almost certainly wrong. (I say "almost certainly" because some psychologists still maintain that Piaget was right, and that the experimental results I shall describe admit alternative conclusions.)

In the 1940s and 1950s, Piaget developed a "constructivist" view of child development. According to this view, a newborn baby enters the world with a cognitive clean slate and, by observing the world around it, gradually pieces together a coherent and steadily increasing understanding of that world. In other words, the child *constructs* a mental model or conceptualization of the world.

Piaget did not arrive at his conclusions by armchair speculation. He was an experimentalist, and his experiments are one reason why his work was so influential. It took great ingenuity—and equipment not available in Piaget's time—for subsequent generations to devise more reliable experiments. When they did so, they reached very different conclusions.

For example, according to Piaget, children younger than ten months old have no proper sense of physical objects as things that endure in the world. Piaget based this conclusion on his observation that, when an object such as a toy is hidden under a cloth, a baby ten months old or younger will fail to reach for it. According to Piaget, "object permanency," as he called it, is not innate but is acquired sometime after ten months of age.

Similarly, Piaget believed that children do not have a number sense until they acquire it at around four or five years of age. In one of Piaget's experiments, repeated many times by different groups, a psychologist would show a four-year-old child two equally spaced rows of six glasses and six bottles and ask whether there were more glasses or more bottles. The child invariably answered that there were the same number. Presumably the child observed a one-to-one correspondence between the rows. The experimenter then spread out the glasses to form a longer row and asked the child again whether there were more glasses or more bottles. Now the child would answer that there were more glasses, apparently misled by the longer length of that row. "Obviously," Piaget concluded, "this shows that the child does not have a properly developed number sense." In particular, Piaget claimed, four- and five-year-old children have not yet grasped the idea of number conservation—the notion that rearranging the objects in a collection does not change their number.

At the time, Piaget's experiments were held up as triumphs of experimental science in psychology. As a pioneer, Piaget was blazing a trail for future generations. And that is good science. Unfortunately, his methods had serious flaws. He relied on the motor actions of the babies in the object permanency test and on a dialogue between the experimenter and the subject for the various number tests performed on older children.

In the case of object permanency, a baby's failure to reach for an object hidden under a blanket does not support the rather dramatic conclusion that the baby thinks the object *has ceased to exist*. Perhaps he simply does not yet have sufficient hand–arm coordination to reach for a hidden object. In fact, we now know that this explanation is correct. Recent experiments, more sophisticated than Piaget's, indicate that even very young babies have a well-developed sense of object permanency.

Likewise, dialogue with a small child is highly unreliable. Communication via language is never 100 percent objective and free of the influences of context, emotion, social factors, and possibly several other things. Just how unreliable dialogue can be was demonstrated by Jacques Mehler and Tom Bever at MIT during the late 1960s.

In one experiment, Mehler and Bever carried out the original Piaget experiment to test for number conservation, but with two- and three-year-old children instead of Piaget's four- and five-year-olds. The children succeeded perfectly. Consequently, unless we believe that children temporarily lose their sense of number conservation between the ages of four and six, we clearly need some alternative explanation for Piaget's results. One is readily available.

Around five years of age, children begin to develop the ability to reason about another person's thought process ("What Daddy means by this is . . . "). This provides the most likely explanation of Piaget's observations. Remember the way the experiment was performed. First the experimenter arranges the glasses and bottles in two equally spaced rows and asks the child which row has more objects. Then the experimenter rearranges one of the rows, making it longer, and again asks the child, "Which row has more objects?"

Now, by four or five years of age, a young child knows that adults are powerful and are knowledgeable. Moreover, she has probably observed the respect her parent showed the experimenter when they arrived at the laboratory. How is this child likely to react when she sees the experimenter rearrange the objects in one of the two rows and then ask *the very same question* as a moment earlier, "Which row has more objects?" She may well reason, "Hmm. That's the same question she just asked me. Adults are not dumb, and this is a special kind of adult who knows a lot. We can both see that the number of objects hasn't changed. So I must have misunderstood the question the last time. I thought she was asking me about the number of objects in the row, but obviously she was really asking me about the length, since that's what she just changed." And so the child gives the answer she thinks is expected of her.

Of course, we can't know for sure. Attempts to find out by interrogating the child are unlikely to yield conclusive evidence, for the same reason that the original Piaget experiment is suspect! This is where the Mehler and Bever experiment came into its own. The kind of "what-does-she-really-want?" reasoning just described is beyond two- or three-year-olds. Mehler and Bever's younger subjects took the experimenters' questions literally, and counted correctly.

What Piaget's original experiment really showed is that four- and five-year-old children can reason rationally about the motivations and expectations of another person. That's an important and useful discovery. But it's not the one Piaget thought he had made!

To confirm that children from age two upward have a good sense of number, Mehler and Bever redesigned the Piaget test to avoid the reliance on language. Their idea was breathtakingly simple. Instead of glasses and bottles they presented the child with two rows of M&Ms. One row contained six M&Ms, the other had four. Sometimes the rows were the same length; sometimes the row of six M&Ms was longer; other times the row of four M&Ms was longer. Instead of being asked to indicate which row had more candies, the child was simply told he could pick one row and eat them. The outcome was precisely what any parent would predict. The child invariably plumped for the row of six candies, regardless of its length. He knew full well which row had more members, and moreover realized that the number was not dependent on the arrangement. The result was just as conclusive with two-year-old children as with four-year-olds.

Another ingenious variation of the original Piaget experiment reached the same conclusion. This time, James McGarrigle and Margaret Donaldson of the University of Edinburgh carried out their experiment in a small puppet theater. Like Piaget, they started by aligning two rows of the same number of objects and asking the child which row had more objects. After the child responded correctly, the experimenter pretended to look away while a teddy bear puppet lengthened one of the rows. Turning back, the experimenter exclaimed, "Oh dear, that silly teddy has mixed up the rows. Can you tell me which row has more objects again?" Children from two to five invariably gave the correct answer. Since the teddy bear had rearranged one of the rows, unseen by the experimenter, the child presumably found it reasonable for the adult to ask the same question again. Yet when the experimenter repeated the process with the same children but rearranged the objects him- or herself, the four- and five-year-old children responded exactly as they had for Piaget, basing their answer on length.

CAN BABIES COUNT?

Many studies have shown that Piaget's conclusions about number sense were wrong. Children as young as two have a well-developed sense of number and of number conservation. What about the underlying constructivist hypothesis, that such numerical abilities are acquired through observation? Perhaps the child learns such abilities in its first two years.

A number of remarkable experiments carried out during the 1980s and 1990s show that's not true either. The first of these experiments was performed in 1980 by Prentice Starkey and his colleagues at the University of Pennsylvania. Starkey's subjects were babies aged between sixteen and thirty weeks. The challenge was to find a way to detect what such young subjects were thinking.

The experimenters relied on measuring the baby's attention span. As any parent knows, any novelty, such as a new toy, will capture a young child's attention for a while. Then that attention starts to wane, or some other novelty takes over. Using video technology, Starkey and his colleagues were able to track the child's attention by recording its gaze. The baby was generally held by its mother on her lap, facing the experimental apparatus.

In one experiment, the baby was shown slides projected onto a screen. A slide was displayed showing two dots side by side. When this first appeared, it caught the baby's attention, and the baby gazed at it for some time. When the baby's attention began to fade and its eyes started to wander, the slide was replaced with a new one that differed slightly from the first. The baby glanced back briefly. The slide was changed again. Each new slide was a slight variation of the previous one. With each repetition, the baby's rearoused gaze—as timed from the videotape—became more and more brief. Then, without warning, the slide showed three dots, not two. Immediately, the baby's interest was aroused, and it gazed longer at the picture (from 1.9 seconds to 2.5 seconds in one run of the experiment). The child clearly detected the change from two dots to three. Another group of babies shown the slides in reverse order noticed the change from three dots to two.

Subsequently, Sue Ellen Antell and Daniel Keating of the University of Maryland used the same method to show that babies just a few days after birth can also discriminate between two and three.

Incidentally, by showing the dots in different arrangements or replacing them with pictures of objects in different configurations, both groups of experimenters eliminated any possibility that some feature other than number was catching the babies' attention. Regardless of what objects were shown or their configuration, the change in *number* was what caught the babies' interest. The conclusion seems definite: babies as young as a few days old can distinguish between collections of two and three objects.

In another experiment, performed by Ranka Bijeljac and her colleagues at the Laboratory for Cognitive Science and Psycholinguistics in Paris, babies just four days old were subjected to auditory stimuli instead of visual ones. Since the stimuli were not visual, the babies' attention could not be measured by timing their gaze. Instead, the experimenters made use of the babies' sucking reflexes to measure their attention. Each baby was given a nipple, connected to a pressure transducer and thence to a computer. When the baby sucked on the nipple, the computer delivered a nonsense word of a fixed number of syllables. Typical three-syllable words might be "akiba" or "bugaloo."

At first, each baby showed great interest in the fact that its sucking produced sounds, and it sucked vigorously. After a while, however, interest waned and the sucking rate dropped. The computer, detecting this drop, at once shifted to delivering words with only two syllables. The baby at once started to suck vigorously again.

Changes from one three-syllable word to another or from one two-syllable word to another did not produce the same increased interest in the baby. Nor did changes in the speed of delivery of each word. It had to be the number of syllables that generated the response.

The Paris experiment shows more than that four-day-old babies can distinguish two from three. It also shows that, even at that age, babies can recognize a syllabic structure in a stream of sound and, in an innate and subconscious way, can detect differences in the number of syllables.

Given that even very young babies can discriminate between two objects and three, does this number sense extend to a sense of (abstract) twoness and threeness? To test this, one would have to show that a baby can see a similarity between collections of, say, two apples and two dots, or three marbles and three bell rings. But how can this be done with a young baby that has no power of speech? Remarkably, ten years ago Starkey and his colleagues found a way to perform just such a test.

The researchers sat babies aged six to eight months in front of two slide projectors. The projector on the left showed a picture of three objects, randomly arranged; the projector on the right showed a picture of two objects, also arranged randomly. At the same time, a loudspeaker situated between the two screens played a sequence of drumbeats. A hidden video camera recorded the baby's eyes to measure its attention.

Initially, each baby examined both pictures, spending a little more time on the picture with three objects, because it was more complex. After the first few trials, however, a remarkable pattern emerged. The baby spent more time looking at the picture whose number matched that of the sequence of drumbeats it heard. When two drumbeats were played, the baby spent more time looking at the picture with two objects, but when three drumbeats were played, it paid more attention to the picture showing three objects.

Starkey and his fellow researchers did not suggest that their subjects were at all conscious of numerosity. The babies' behavior was probably an in-built neuronal response, in which activation of a certain pattern of neuronal firing caused by hearing two sounds caused preferred receptivity to scenes showing two objects. That does not amount to a sophisticated conception of number. But it is most definitely a number concept. And it tells us that however we may rate our mathematical ability, we all have a built-in number sense. We were born with it.

Perhaps we were also born with the ability to add—at least to the limited extent that chimpanzees can add.

BUT CAN THEY ADD?

Can babies add? Does a baby know that 1 + 1 = 2 or that 2 + 1 = 3? In 1972 the American psychologist Karen Wynn stunned the world when she announced her findings that babies as young as four months were capable of performing (in an innate, non-conscious fashion) simple additions and subtractions. As with all experiments performed on babies, the main challenge was to find a way to discover what was going on in the mind of so young a subject.

Wynn made use of babies' sense of "the way the world is." Even very young babies are troubled when they encounter something that runs counter to the laws of physics. For example, an object apparently suspended in mid-air with nothing visible to support it will elicit an intent stare.

Wynn sat her young subjects in front of a small puppet theater. The experimenter's hand came out from one side and placed a Mickey Mouse puppet onto the stage. Then a screen came up, hiding the puppet. The experimenter's hand appeared again, holding a second Mickey Mouse, which it placed behind the screen. Then the screen was lowered. Sometimes the baby would see two Mickey Mouses on the stage. On other occasions, the lowering of the screen would reveal just one Mickey Mouse, the other having been removed unseen through a trapdoor. By videotaping the baby's responses, Wynn was able to measure the time the baby gazed at the stage. On average, when the lowering of the screen revealed one puppet rather than two, the babies looked for a full second longer.

The most obvious conclusion is that, having seen two puppets placed on the stage, one after the other, and none removed, they expected to see two puppets at the end. They "knew" that 1 + 1 = 2 and were surprised to see a manifestation of the incorrect rule 1 + 1 = 1.

The subjects' attention times were also longer when the lowering of the screen revealed three puppets. Apparently, they "knew" that 1 + 1 was equal to neither 1 nor 3, but exactly 2.

The babies also seemed to know subtraction. To test this, the demonstration started with two puppets on the stage, then the screen went up, and then the baby saw one puppet removed. When the lowering of the screen revealed two puppets (illustrating $2 - 1 = 2$), the babies gazed at the stage by as much as three seconds longer than when only one was left (illustrating $2 - 1 = 1$).

To eliminate the possibility that the babies in Wynn's experiment were relying on a visual memory rather than a numerical one, the French psychologist Etienne Koechlin repeated the procedure, but placed the objects on a slowly revolving turntable. The constant movement on the stage prevented the babies from forming a fixed mental image of what they saw; they could not predict what the scene would look like when the screen was lowered. Nevertheless, they showed far greater attention when the number of objects on the turntable was not what it should have been.

Interestingly, of all the features of the physical world that babies are born with, number seems to rank among the most significant—far more so than physical form or appearance. This was demonstrated by the American psychologist Tony Simon and his colleagues in the early 1990s, using yet another variation of Wynn's experiment. If a baby sees two puppets disappear behind the screen, it shows no surprise when the screen is lowered to reveal, say, two red balls; but it appears troubled when just one ball is revealed. Apparently the idea that one object can transform into another is less troubling than is a change in number. This provides a new and totally unexpected twist to Piaget's concept of object conservation.

Further confirmation of this odd view of object conservation is provided by the outcome of another experiment, in which a baby less than twelve months old is placed in front of a screen from behind which, say, a red ball and a blue rattle alternately appear. Provided the baby does not see the two objects at the same time, it will not show surprise to see just one object (either a ball or a rattle) when the screen is lowered. It appears happy to accept that there is a single object that sometimes looks like a red ball and at other times like a blue rattle. Only when it is a year old or more do the two different appearances begin to imply the existence of two objects. In a baby's first year of life, number is apparently a more important "invariable" than color, shape, or appearance.

Incidentally, these remarkable arithmetical abilities of young babies are strictly limited to simple additions and subtractions involving the numbers 1, 2, and 3. Babies younger than one year seem unable to distinguish four objects from five or six.

In the next chapter, we shall turn to the uniquely adult-human numerical world beyond 3. But before then, spare a thought for the very small number of individuals who have to live in a number-dependent society without having any sense of what numbers are.

BUTTERWORTH'S CASEBOOK

The British psychologist Brian Butterworth, who studies brain abnormalities, has collected a number of fascinating cases of individuals who have lost or never had any sense of number. He describes some of them in his 1999 book *The Mathematical Brain* (published in the United States as *What Counts*).

For instance, one of Butterworth's examples, Signora Gaddi, suffered a stroke that left both her language and reasoning abilities intact, but completely destroyed her numerical capacity. She could not determine or estimate the number of objects in any collection. Moreover, she could only recite the number words up to four and, as a result, could only count the members of a collection of four or fewer objects.

Or consider Frau Huber, who had an operation to remove a tumor in her left parietal lobe. After the surgery, her general intelligence and language ability seemed just fine, but numbers literally had no meaning for her. She could not even be taught finger addition. She could recite the multiplication table, but it was just a "nonsense poem" to her. Although she was able to learn new arithmetical facts verbally, they had no meaning for her and she could not make any use of them. She was unable to work out any arithmetical fact.

Prior to her surgery, Frau Huber did have a number sense; but some individuals are born without it and are never able to acquire it. Another of Butterworth's examples, whom he refers to as Charles, is a highly intelligent young man who earned a degree in psychology. But Charles has virtually no sense of numbers. Faced with simple arithmetic, his only recourse is to use his fingers. To perform any kind of calculation, he needs a calculator. The answer he gets means nothing to him. He is unable to tell which of two numbers is the larger. Forced to make such a comparison, he has to resort to counting to see which number he reaches first. If Charles is shown a collection of, say, three objects, he is unable to say how many there are, but must count them. When asked to add or subtract two numbers, he does so by counting on, but his performance is slow and erratic. In one test, he took eight seconds to add 8 and 6, and twelve seconds to subtract 2 from 6. He failed on $7 + 5$ and on $9 + 4$. Not surprisingly, it took Charles longer than usual to earn his degree.

Another highly intelligent individual having virtually no number sense is Julia. When Butterworth first examined her, she had completed a first university degree and was embarking on a postgraduate program. She could perform simple arithmetic only by using her fingers and was helpless when faced with numbers beyond the range of her hands. She had no sense of fractions or of the basic arithmetical properties—for example, she did not know that $3 \times (2 + 5)$ is the same as $(3 \times 2) + (3 \times 5)$. She was able to count 1, 2, 3, etc., but could not count in threes except by counting in ones and stressing every third number: one, two, *three*, four, five, *six*, seven,... Unlike Charles, however, she was able to compare two numbers to see which was the larger.

People like Charles and Julia suggest that number sense is not something that can be learned. Despite having the intelligence to earn college degrees, they were unable to acquire a sense of number. They could learn facts about numbers, but those facts had no meaning for them.

·3·

EVERYBODY COUNTS

THE NUMBER SENSE built into newborn babies is similar to that observed in rats, chimpanzees, and monkeys. In fact, Mark Hauser of Harvard University and his colleagues have carried out Karen Wynn's original experiments with rhesus monkeys, obtaining similar results. But among all the animal species, only humans seem able to take this deeply ingrained but strictly limited ability and extend it to the point where we can quantify and talk about collections ranging into the billions. Do we in fact extend our basic, built-in number sense, or do these larger collections call on an entirely different mental ability?

Certainly our brains appear to handle collections of three or fewer members differently from the way we deal with larger collections. In experiments where adult subjects are asked to name the number of dots randomly arranged on a slide shown to them, the time required to give an answer is almost identical for one and two dots, and only slightly longer for three (just over half a second). Beyond three, however, the time starts to increase rapidly. As the number of dots grows, so too does the number of errors.

The sudden change in behavior beyond three objects suggests that the brain may be using two different mechanisms. For collections of three or fewer objects, recognition of numerosity appears to be virtually instantaneous, and achieved without counting. For collections of four or more objects, however, it is at least plausible that the result is obtained by counting. The time required to give the number of objects in the collection increases linearly as the number goes up from three to six. This is exactly what you would expect if the answer was arrived at by counting.

Further confirmation that the brain generally handles collections of three or fewer by an "all at once," instinctive (and subconscious) process comes from studies of patients with certain cerebral lesions. Although brain damage often affects large parts of the brain, destroying many mental faculties, sometimes a lesion is very local and has just one or two very specific effects. Patients with such "focused" lesions provide cognitive psychologists with extremely useful evidence they would otherwise be unable to obtain. In one case, described by Dehaene in *The Number Sense,* a patient in Paris suffered a brain lesion that destroyed her ability to count, or even to check off objects one after another. Yet when three or fewer dots were flashed onto a computer screen, she could immediately report the correct number of dots.

But even if we use a different mechanism to arrive at the numerosity of collections of more than three members, the human concept of number is almost certainly built upon our innate sense of number as being *a property of collections of discrete physical objects.* For instance, if you show a three- or four-year-old child two red apples and three yellow bananas and ask her how many different colors there are or how many different kinds of fruit she can see, she will answer "five." The correct answer is two, of course. But so deeply ingrained is the idea that number is a property of collections of discrete physical objects that it takes the relative sophistication of five years' experience in the world before a child can overcome that built-in notion and apply numbers in a more abstract fashion.

What happens when we are faced with a collection of considerably more than three objects? Are we able to estimate how many there are? Can we distinguish between two collections having different numbers of members?

Certainly we know that if we are asked to give the exact number of objects in a collection having, say, ten or more members, our only strategy is to count them, either one by one or perhaps two by two or even three by three. But what if we just want a reasonably good estimate? How good are adult humans at estimating the number of objects in a collection of, say, a hundred?

The experimental evidence shows that we are better than you might think, although we can be misled by surrounding circumstances. For example, when asked to estimate the number of dots on a page, subjects will typically underestimate the total if the dots are placed irregularly and overestimate if they are in a regular pattern. Again, if shown twenty-five or thirty dots on a page, we will overestimate the total if the collection is surrounded by a border of hundreds of other dots and underestimate if the border contains just ten or twenty dots.

When not misled by surrounding circumstances, however, we do seem capable of providing good estimates of collection sizes, and people who regularly need to estimate say, large crowds of people can become quite good at it. Moreover, our ability increases with just a small amount of mental priming. For instance, if we are first shown a collection that we know contains, say, 200 dots, we can provide pretty good estimates of collections of between 10 and 400 dots.

Easier than estimating the total is spotting that two collections have different sizes. Here a distance effect comes into play. It is much easier to distinguish the sizes of collections where the difference between them is significant compared to the totals than when the difference is relatively small. For example, we can generally tell that collections of, say, 80 and

100 objects have different sizes, but are unable to spot the larger of two collections of 80 and 85 objects. In fact, our ability to distinguish collection sizes is described by a remarkably precise numerical law, called "Weber's law" after the German psychologist who discovered it. If a subject can discriminate, say, a set of 13 dots from a reference set of 10 dots with a success rate of 90 percent, then the same subject will be able to discriminate 26 dots from a reference set of 20 with the same degree of accuracy. As long as the ratio of the test set to the reference set is held constant, the subject's success rate will remain constant as well. It's as precise a linear relationship as you are likely to find in psychology!

The Weber effect has also been observed in monkeys. David Washburn and Duane Rumbaugh tested two monkeys, Abel and Baker, who had previously been taught the numerals 0 to 9—in the sense that they could associate the numeral with a collection of objects having that many members. The monkeys were trained to use a computer joystick to select one of a pair of numerals presented on the computer screen. They were rewarded with as many small pellets of food as the number they had selected, thereby encouraging them to select the larger number. After a training period, Washburn and Rumbaugh measured their performance. The further apart the two numbers, the faster was their response time and the greater their accuracy.

So there is some similarity between monkeys and humans, but not much. When it comes to estimating sizes of collections and detecting differences in collection sizes, adult humans are uniquely able to extend the number sense far beyond 3.

Moreover, humans can do much more than that. In addition to our innate number *sense*, which gives us reasonably good estimates of the sizes of collections, we have counting, which gives us exact totals. We also have arithmetic, which gives us precise numerical answers to a whole range of questions. Where did these unique features come from? Were we born with these abilities or are they skills we learn?

THE COUNTER CULTURE

Two keys unlock the door to the numerical world beyond 3, and as far as we know, only our species has ever found either one. The first is the ability to count. The second is the use of arbitrary symbols to denote numbers and thus manipulate numbers by the (linguistic!) manipulation of those symbols. These two attributes enable us to take the first step from an innate number *sense* to the vast and powerful world of mathematics. Both abilities, counting and symbolic representation, are among the collection of (related) abilities that our ancestors acquired between 75,000 and 200,000 years ago, as we shall discover in Chapter 8.

Counting is not the same as saying how many members are in a collection. The number of members in a collection is just a *fact* about that collection. Counting those members, on the other hand, is a *process* that involves ordering the collection in some fashion and then going through it in that order, counting off the members one by one. (I shall ignore variations where the collection is counted by twos or threes. Those are just that: variations.) Since counting does in fact tell us the number of members in a collection, we often confuse the two. But that is a consequence of familiarity. Very young children view counting and number as quite unconnected. Ask a three-year-old to count up his toys and he will perform flawlessly: "One, two, three, four, five, six, seven." He may well point to each toy in turn as he counts. But now ask him how many toys he has, and chances are he will tell you the first number that pops into his head—which may not be seven. At that age, children simply do not relate the process of counting to that of answering the question "How many?"

Incidentally, even though a three-year-old will probably not know how many objects make "three," a two-and-a-half-year-old does realize that number words are different from other adjectives. For instance, suppose you show a two-and-a-half-year-old a picture of a single red sheep and another of three or four blue sheep. If you say "Show me the red sheep," the child will point to the red sheep. Say "Show me the three sheep" and the child will point to the picture of the blue sheep. The latter may have three or four sheep; the child does not appear to know exactly what number "three" is. But she does know that "three" applies to a collection of objects, not to a single object. She also knows that it is okay to say "three little sheep" but not say "little three sheep." In short, by age two and a half, children realize that number words are different from other words.

It is around the age of four that children realize that counting provides a means to discover "How many?" Part of that realization must be the recognition that, when you are counting the members of a collection, *the order in which you count doesn't matter.* Regardless of which object is "first," "second," and so on, the number you finally reach is always the same. With this remarkable insight, which we all simply take for granted, we enter the counter culture.

My use of the word "culture" just then was more than a cute pun. There are a few cultures today that do not have counting (at least, they do not have counting beyond two, which arguably means they don't have counting at all). In the past there were many more. Most likely our ancestors did not have counting either. Particularly suggestive evidence is that the words we use for the first three numbers—the ones that correspond to our innate number sense—are very different from all the others.

For example, starting with four, there is a simple rule for turning the number word into the corresponding adjective for use in an ordered list: add the suffix "th" (and possibly make some minor change to facilitate pronunciation). Thus, "four" becomes "fourth", "five" becomes "fifth", "six" becomes "sixth", and so forth. For the first three numbers, however, there are completely different words for the ordinals: "first", "second", and "third". Moreover, these words have other, related uses. "Second" can be used as a verb, meaning "to support", and we have the adjective "secondary". The Indo-European root of the word "three" is related to the Latin prefix *trans* (meaning *beyond*), the French word *trés* (meaning *very*), the Italian *troppo* (meaning *too much*), and the English *through*, all of which suggests that it was once the largest numeral.

The other European languages likewise show special treatment of the first two numbers followed by a more regular form for all the others. In French we have: *un/premier*, *deux/second* (or *deuxième*), and then the more regular *trois/troisième*, *quatre/quatrième*, etc. In German we find: *ein/erst*, *zwei/zweite* (or *ander*), and then *drei/dritte*, *vier/vierte*, etc. In Italian: *uno/primo*, *due/secondo*, and then *tre/terzo*, *quattro/quarto*, etc.

Further, we have special ways to talk about collections of two or three objects. For two, we speak of a pair, a brace, a yoke, a couple, or a duo, and we have the adjective "double." For three, we have the words "triple," "trio," and "treble," Beyond three, however, we make use of the regularly constructed forms "quartet," "quintet," "sextet," etc. Many of the words for a collection of two objects are restricted to certain kinds of objects, consistent with the idea that the innate numbers one, two, and three are intimately connected with collections of physical objects. Thus, for example, we speak of a brace of pheasants, a yoke of oxen, and a pair of shoes, but cannot talk about a "brace of shoes" or a "yoke of pheasants."

How and when did our ancestors first develop the idea of counting, as opposed to estimating using their innate number sense? Quite possibly they started out dealing with collections the way the Aboriginal tribe the Warlpiris do today, by considering just three possibilities: *one*, *two*, and *many*, with three being the point where exact counting ends and the collection is simply "large."

But how did our ancestors learn to count beyond three?

Perhaps they began the way young children do today. As any parent or elementary school teacher knows, children learning arithmetic spontaneously use their fingers. Indeed, so strong is the urge to count on their fingers that if a parent or teacher tries to insist that the child do it the "right" or "grown-up" way, the child will simply use her fingers surreptitiously. As for the idea that dispensing with fingers is the "adult" way, many adults also do arithmetic with their fingers.

Certainly, our base-10 number system is evidence that counting began as finger-enumeration. Since we have ten fingers, if we use our fingers to count, we run out of counters when we reach ten, and we have to find some way of recording that fact (perhaps by moving a pebble with our foot) and then starting again. In other words, finger (and thumb) arithmetic is base-10 arithmetic, where we "carry" when we reach ten. (The satirist and mathematician Tom Lehrer once quipped that base 8 is just the same as base 10 "if you're missing two fingers.")

Further evidence in favor of the hypothesis that arithmetic began with finger manipulation is our use of the Latin word *digit* to mean both *finger* and *numeral*.

Admittedly, neither piece of evidence is overwhelming, but they are suggestive. The same can be said for my next piece of evidence, this time from the neuroscience laboratory.

By various techniques scientists are able to measure the level of activity in different parts of the brain while it is engaged in a particular task. For instance, the massive frontal lobe is where most brain activity takes place when a person is using language. In a sense, the frontal lobe is the brain's "language center."[1]

1 A word of caution is in order here. Although it is convenient and often useful to regard the brain as consisting of several different "modules"—a language module, an arithmetic module, and so on—the brain is a single organ. Strictly speaking, it is the *brain* that is doing the language processing, not the frontal lobe.

It is a bit like an automobile engine. When the engine is running, the actual source of power is in the combustion cylinders, but the whole engine is required in order to function, and it is the *engine* that provides the driving force. The statement that the frontal lobe is the brain's "language center" means that this part of the brain is most intensely active during language processing. Moreover, damage to this part of the brain is likely to curtail or destroy linguistic ability. But strictly speaking, it is the whole brain that handles language.

Personally, I generally regard talk of distinct "modules" of the brain as largely metaphorical (though many authorities seem to use the word literally). Just because damage to a particular part of the brain causes loss of a certain mental faculty, say arithmetic, does not mean that other parts of the brain are not involved in arithmetic, only that the part in question was (formerly) essential to the performance of arithmetic. In some cases, other parts of the brain will eventually take over from the damaged part, sometimes leading to a complete recovery of the lost faculty.

When a typical person is performing arithmetic, the most intense brain activity is in the left parietal lobe, the part of the brain that lies behind the frontal lobe. Now, as it happens, similar studies have shown that the left parietal lobe is also the region that controls the fingers. (It requires a considerable amount of brain power to provide the versatility and coordination of our fingers, and hence a large part of the brain is devoted to that task.)

The question leaps to the mind at once: is it a coincidence that the part of the brain we use for counting is the very part that controls our fingers? Or is it a consequence of the fact that counting began (with our early ancestors) as finger-enumeration, and that over time, the human brain acquired the ability to "disconnect" the fingers and do the counting without physically manipulating them?

Clinical psychologists have also found a connection between finger control and numerical ability. Patients who sustain damage to the left parietal lobe often exhibit Gerstmann's syndrome, a lack of awareness of individual fingers. For instance, if you touch one finger of a typical sufferer of Gerstmann's syndrome, the individual cannot say which finger you touched. These people are also typically unable to distinguish left from right. More interestingly, from our point of view, people with Gerstmann's syndrome invariably have difficulty with numbers. For example, both Frau Huber and Signora Gaddi, the individuals having no number sense whom we met in the previous chapter, suffered from Gerstmann's syndrome.

It is tempting, to say the least, to speculate as to the reason for the correlations between Gerstmann's syndrome, finger control, and numerical ability. If early *Homo sapiens*' first entry to the world of numbers was via their fingers, then the large region of the brain that controls the fingers would be the one in which their descendants' more abstract mental arithmetic would be located. Very likely, our present-day, strictly mental number sense is an abstraction from our ancestors' physical finger manipulations. Mental arithmetic may be in essence "off-line" finger manipulation, which became possible when the human brain became able to disconnect the brain processes associated with finger manipulation from the muscles that control the fingers.

It's a possibility. An alternative early means of counting—which could have co-existed with finger counting—was to use some sort of physical tally system such as making notches on a stick or a bone. Notched bones have been discovered that date back at least 35,000 years. Thus, our ancestors may have used a physical enumeration system as much as 40,000 years ago, the time when, according to the fossil record, humans started to use symbolic representations in cave paintings and rock carvings.

Of course, use of fingers or tallies indicates a conception of numerosity, but does not necessarily imply a concept of *number,* which is purely abstract. Abstract numbers are the key to modern mathematics. How and when did we acquire them? The best current evidence we have for the introduction of *abstract* counting numbers, as opposed to markings, was discovered by the University of Texas archaeologist Denise Schmandt-Besserat in the 1970s and 80s. At archaeological sites in Iraq, where a highly advanced Sumerian society flourished from around 8000 to 3000 BC, Schmandt-Besserat kept finding small clay tokens of different shapes, including spheres, disks, cones, tetrahedra, ovoids, cylinders, triangles, and rectangles. The older ones were simple, later ones often quite intricate. Gradually, as she and other archaeologists slowly pieced together a coherent picture of Sumerian civilization, Schmandt-Besserat realized that the puzzling clay tokens were counters used to support commerce. Each shape represented a certain quantity of a particular item—metal, a jar of oil, a loaf of bread, and so on.

The first tokens appeared around 8000 BC. By around 6000 BC they had spread throughout the Fertile Crescent, which stretched from present-day Iran to Syria. Over the years, even more elaborate shapes were produced, including parabolas, rhomboids, and bent coils, and some of the tokens began to carry markings.

Two ways were developed to keep a complete record of a man's possessions. The more elaborate markers were strung like beads across an oblong clay frame. The simpler tokens were wrapped in a flat sheet of wet clay, which was then sealed and allowed to harden into a spherical envelope. Both methods were fully functional early forerunners of our present-day bank accounts. But the clay envelopes had an obvious drawback. Whenever someone wanted to trade, or even check the status of his account, the envelope had to be broken and a fresh one made. To reduce the frequency of envelope breaking, the Sumerians hit upon the idea of impressing on the wet clay of the envelope each token that was to be enclosed, prior to the envelope being sealed. That way, the outer surface of the envelope carried an indelible record of what was inside.

In time the Sumerians realized that with these visible records on the outer surface of the envelope, the contents were unnecessary. The imprints of clay tokens could represent possessions. So the Sumerians stopped sealing the tokens in clay envelopes and instead simply impressed the tokens onto a flat tablet of wet clay.

The step from impressed clay envelope containing tokens to impressed clay tablet is the first known use of abstract markings to denote counting numbers. In fact, the Sumerian accounting tablets are the earliest known writing system, which means that the use of markings to denote numbers preceded the use of markings to denote words. To put it another way, early Shakespeares obtained the tools of their trade from early Newtons.

Of course, since the Sumerians had different tokens, and correspondingly different markings on the clay tablets, to represent quantities of different kinds of items, they did not really have a single number system. Rather, each kind of item had its own separate number system. But by eliminating the need for physical tokens and using symbolic markings instead, they made the first major step toward the all-purpose, abstract numbers we use today.

SYMBOLS OF SUCCESS

One of the things that makes arithmetic possible—for those who can do it!—is the extremely efficient symbolic notation we have for representing numbers. For the Romans, with their cumbersome notation of Is, Vs, Xs, and so on, even the simplest of arithmetic sums was difficult to calculate. And they had no way to represent fractional or negative quantities.

The number notation we use today is far more efficient. Using just the ten digits 0, 1, 2, 3, 4, 5, 6, 7, 8, and 9, we can represent any positive whole number. The trick is to use these digits to form numerical "words" that name numbers, just as we put letters together to create words that name various objects or actions in the world. The number denoted by a particular digit in any numerical "word" depends on its position in that word. Thus, in the number

1492

the first digit 1 denotes the number *one*-thousand, the second digit 4 denotes *four*-hundred, the third digit 9 denotes *nine*-tens (or ninety), and the last digit 2 denotes the number *two*. The entire "word" 1492 denotes the number:

one-thousand and *four*-hundred and *nine*-ty *two*

Our present number system was developed over two thousand years by the Hindus, reached essentially its present form in the sixth century, and was introduced into the West by Arabic mathematicians in the seventh century. As a result, it is generally called the "Arabic system." It is almost certainly the most successful conceptual invention of all time and is the only genuinely universal language on earth. It has gained worldwide acceptance because it is far better designed and much more efficient for human usage than any other system.

One of the strengths of the Arabic system is that the number "words" can be read aloud, and moreover, the spoken version reflects the numeric structure in terms of units, multiples of ten, multiples of a hundred, and so forth. (Later we shall consider the arithmetical consequences of the variations in reading Arabic numbers in different languages.)

Another strength of the Arabic system is that it *is* a language.

Consequently, it allows humans—who have an innate linguistic fluency—to use their language ability to handle numbers. Thus, while our intuitive number sense resides in the left parietal lobe, the linguistic representation of exact numbers is handled by the frontal lobe, the language center. I shall have more to say on this later.

Although the use of the symbols 1, 2, 3, 4, 5, 6, 7, 8, 9 for the digits is now universal, in the past there have been others, including Cuneiform, Etruscan, Mayan, ancient Chinese, ancient Indian, and Roman. The Chinese still use a modern variant of their ancient system in addition to Arabic notation, and of course Western societies still use Roman numerals for some specialized purposes.

In the light of our earlier discussions about the special nature of the first three counting numbers, it is of interest to note that in every number representation system ever used, the first three numbers are denoted the same way: 1 is denoted by a single stroke or dot, 2 by placing two such symbols side by side, and 3 by placing three such symbols side by side. In the Roman system, for example, the first three numerals are I, II, III. Mayan notation uses dots: •, ••, •••. The different systems differ only from the fourth numeral onward.

What's that you say? Our Arabic system does not follow that pattern? Yes, it does. The ancient Indian system used horizontal bars, like this: —, =, ≡. When people started to write these symbols without taking the pen from the paper, it yielded this pattern: —, Z, Ǝ. At some stage the first stroke became vertical. When printing was invented, the numerals were given the stylized versions we use today: 1, 2, 3.

Once you have a system for representing any positive whole number, it can be easily extended to fractional and negative quantities. The introduction of the decimal point or the fraction bar allows us to represent any fractional quantity (3.1415 or $^{31}/_{50}$, for example). The minus sign (–) extends the range to all negative quantities, whole or fractional. (Negative numbers were first used by sixth-century Indian mathematicians, who denoted a negative quantity by drawing a circle around the number, but negative numbers were not fully accepted by European mathematicians until the early eighteenth century.)

Incidentally, the original idea of denoting numbers by having a small collection of basic symbols and stringing them together to form number "words" is due to the Babylonians, around 2000 BC. Because it was built on the base 60, the Babylonian system itself was cumbersome to use, and thus did not gain widespread acceptance, although we still use it in geography and our measurement of time (60 minutes make one hour, 60 seconds make one minute).

One of the most powerful aspects of Arabic notation is that arithmetic can be performed through fairly straightforward (and easily learned) manipulations of the symbols. When performing addition, for example, we write all the numbers one beneath the other, aligned in columns starting from the right, and then proceed to add the digits in each column from right to left. Whenever a sum in a column reaches 10, we put a 0 in that column and carry a 1 to the next column to the left. This procedure works for any set of numbers and can be carried out automatically by a machine. For the other basic arithmetical operations, subtraction, multiplication, and division, there is likewise a standard procedure that always works, regardless of what the actual numbers are.

Arabic notation makes basic arithmetic so mindless that in the days before cheap hand-held calculators, elementary arithmetic was one of the least popular classes in schools. It is a great pity that for so many years our teaching methods have obscured one of humankind's greatest conceptual inventions. The true marvel of the invention of the Arabic number system is lost in the trivia of symbolic manipulation. Arithmetic is a dull and mindless task which our creative intellect has found ways to automate. But let's not forget what an incredible human invention is the Arabic number system. It is concise and easily learned. It allows us to represent numbers of unlimited magnitude and apply them to collections and measurements of all kinds. Moreover, its greatest strength is precisely that it reduces computation with numbers to the routine manipulation of symbols on a page.

NUMBERS IN THE MIND

Now, despite the fact that everyone is taught to carry out calculations using Arabic notation, you might think that, apart perhaps from the "math types," most of us attach little significance to what number symbols mean. The more so in view of our observation that our innate number sense does not extend far beyond 3, if at all (except in the sense of making general estimates of collection sizes). But this is not at all the case. Psychologists have known for twenty years that every one of us attaches great significance to each and every counting number we encounter. You may think you are not a "numbers person." But the evidence says otherwise.

For one thing, consider the number comparison test we encountered earlier, in which you were asked to decide which of two numbers is larger (page 17). This test was first carried out by the American psychologists Robert Moyer and Thomas Landauer in 1967. They flashed pairs of digits, such as 7 and 9, onto a computer screen and asked the subject to declare which was the larger by pressing one of two response keys. The responses were timed electronically.

This sounds easy until you try it. Response times of half a second or more are not uncommon. Moreover, under the pressure of the experiment, subjects sometimes make errors. The surprising result that Moyer and Landauer observed was that, for each individual, the response times varied systematically with the relationship between the numbers presented. Given two numbers with one much smaller than the other, say 3 and 9, subjects responded quickly and accurately. If the two numbers were closer, say 5 and 6, response times were 100 milliseconds or more longer, and subjects gave the wrong answer in as many as one out of ten cases. Moreover, for a fixed separation, response times increased as the numbers got larger. Choosing the larger of 1 and 2 was easy, it took a little longer for 2 and 3, and much longer for 8 and 9.

Variations of the Moyer and Landauer experiment have been repeated many times, always with similar results. What's going on? The shapes of the actual symbols cannot account for the difference—the pair of symbols 3 and 8 don't look that much different from 8 and 9, and yet subjects take longer to decide that 9 is greater than 8 than to decide 8 is greater than 3. The critical factor seems to be the number itself, not the symbol.

Further evidence comes when we extend the experiment to two-digit numbers. Try this one. Which is larger, 72 or 69? Got it? All right, here's another. Which is larger, 79 or 63? If you are like most people who have taken this test, two observations will apply. First, it took you longer to decide that 69 is smaller than 72 than to decide that 63 is smaller than 79. The greater distance between the numbers made the second pair easier. Second, you did not answer the questions in the most efficient way, which is to notice that you need not look past the first digit in each pair. *Any* two-digit number starting with 6 will be smaller than *any* two-digit number starting with 7. If you were to answer the question that way, your time should be the same for both examples. But what everybody does is go straight from the symbols to their mental conception of the actual numbers as entire units. This also explains why it takes you only slightly longer to decide which of 67 and 69 is larger than it does for 67 and 71; the fact that the first pair both start with the same digit, whereas the second pair start with different digits, makes hardly any difference. If you based your decision on the symbols rather than the numbers they represented, it would make more of a difference.

Even more dramatic evidence of the way numbers develop a life of their own in our brains was provided in the early 1980s by two Israeli researchers, Avishai Henik and Joseph Tzelgov. They showed subjects pairs of digits in different-sized fonts on a computer screen, and measured the time it took the subject to decide which symbol was printed in the larger font. On the face of it, this task has *nothing* to do with what number the digit denoted, but only with the size of the actual symbol. Nevertheless, subjects took longer to respond when the relative size of the fonts conflicted with the relative sizes of the numbers. For example, it took longer to decide that the symbol **3** is larger than the symbol **8** than to decide that the symbol **8** is larger than the symbol **3**. Subjects were unable to forget that the *number* 8 is larger than the *number* 3. This means that digits are not just symbols to which meaning *can* be attached; they are symbols to which meaning *is* attached, and closely.

The meanings attached to digits and to multi-digit expressions are numbers, of course, but what does that mean? In what sense do we have numbers *in our minds?*

Number comparison experiments and other investigations suggest that we have a sort of "mental number line," where we "see" the numbers as points on a line, with 1 at the left, 2 to its right, then 3, etc. To decide which of two numbers is the larger, we mentally locate them on our mental number line and see which is on the right.

This idea of a mental number line suggests the mathematicians' number line we learn about in elementary school. But there is one significant difference. On our mental number line, the numbers are not spaced evenly apart as they are on the mathematicians' number line. Rather, the farther we go along our mental number line, the closer together the numbers appear to be. This explains the results of Moyer and Landauer's number comparison test. This increasing compression of the numbers makes it more difficult to distinguish the two members of a pair of larger numbers than a pair of smaller numbers. Subjects can decide which is the larger of 5 and 4 much faster than for the pair 53 and 52. Even though the numerical difference between each pair is the same, namely 1, the larger pair seem "closer together" on our mental number line than do the smaller pair.

Moreover, the symbols we use to denote numbers appear to become "hard-wired" into our intuitive number module in the left parietal lobe, in a way that the ordinary language number words (handled by the frontal lobe) do not. In *The Mathematical Brain,* Butterworth gives some clinical evidence to support this apparent distinction.

For instance, Butterworth reports, there are people who are unable to read words but who can read aloud single or multi-digit numbers presented to them using numerals. Conversely, there is a man—Dottore Foppa, an early Alzheimer's sufferer—who can read words, including number words and word expressions for multi-digit numbers, but is unable to read aloud a number of two or more digits presented to him in numerals. An extreme case is presented by Donna, a woman who had surgery to her left frontal lobe. Although she can read and write single- or multi-digit numbers in numerals, not only can she neither read nor write words, she can name only about half the letters of the alphabet. Despite being unable to write her own name—the result is an illegible scribble—on a standard arithmetic test (where the questions are presented in purely numeric form) she does just fine, writing her numerals neatly, in columns, and invariably getting the right answer.

In other words, a number system such as the Arabic system may indeed be a language, but it is a very special one handled in a different region of the brain from normal language. If (as I suggested earlier) our counting ability derived from finger use by our ancestors—a physical process controlled by the same left parietal lobe in which resides our number sense—this distinction between number symbols and number words is just what one would expect if our number symbols derive from the use of our fingers and number words come from ordinary language.

Incidentally, some individuals who are unable to compare pairs of numbers "by inspection" using their number sense can nevertheless tell which of two numbers is the larger by counting—they start to count 1, 2, 3, etc., and see which number they reach first. Since memorization of the number sequence is a linguistic feat, this strategy would appear to make use of the brain's language faculty to make up for a lack of number sense.

Similarly, some people who cannot add or subtract pairs of fairly small numbers from memory use a counting-on strategy. To add 5 and 4 they will proceed thus: "Five, six, seven, eight, *nine*—the answer is nine." They may recite the numbers out loud, or in their heads, and may use their fingers to keep track of how many words they say. This approach is more efficient if the larger of the given pair of numbers is taken as the starting point for the count, so individuals who can compare pairs of numbers have an advantage over people who cannot. To calculate 6 – 3 they will recite (internally or out loud, and often counting the words on their fingers): "three, *four, five, six*—the answer is three."

The idea of a mental number line also explains the following phenomenon. (To get the proper effect, it is important that you follow my instructions exactly, in the order I give them. Don't look ahead to see what's coming next.)

Memorize the following list of digits: 7, 9, 6, 8.

Cover up the list with your hand.

Now count down from 16 to 1 in threes. Keep the list covered up.

Have you done that?

Keep the list covered up until you have finished counting.

Now tell me whether 5 was in the list you just remembered. Was 1 in the list?

Now you can uncover the list.

Almost certainly, you felt pretty confident that 1 was not in the list, but less certain about 5. The explanation is that you formed a mental image of the list as consisting of numbers in the region of 6 and 7 on your mental number line. When later asked if 5 was in the list, you were not sure, because 5 is in the general region of the list. But because 1 is well away from that region, you were confident it was not in the list.

Only about 14 percent of adults say they are conscious of having a mental number line. Of these, most of those whose primary language reads left to right claim that their number line runs left to right, although some say theirs goes upward, and a few see it running downward. However, in an ingenious experiment performed in 1993, Dehaene and his colleagues showed that most of us have a mental number line, even if we are not aware of it, and that we compare pairs of numbers by comparing their positions on our mental number line.

Dehaene presented subjects with a series of numbers on a computer screen, which they had to categorize as even or odd, as rapidly as possible, by pressing one of two buttons. For half the subjects, the "odd" button was in their left hand, the "even" button in their right; for the other half, they were reversed. The computer timed the subjects' responses. The outcome was amazing. With small numbers, subjects responded faster with their left hand; with larger numbers, they were faster with their right hand. Dehaene's explanation for this difference is that the subjects "saw" the numbers arranged spatially on a left-to-right line. Smaller numbers were at the left of this line, so the subjects responded faster with their left hand; larger numbers were on the right, and this led to a faster response with the right hand.

Interestingly, there were some Iranian students in the group tested, and their responses were reversed—they were faster with their right hands for smaller numbers and faster with their left hands for larger numbers. Iranians read right to left, of course. Thus it appears that our mental number line generally runs in the direction of our reading.

THE SOUND OF NUMBER

A few years ago I attended a luncheon at which the mathematician Arthur Benjamin was giving a demonstration of his amazing feats of mental arithmetic. Just before he was due to begin, he asked to have the air conditioning turned off. While we were waiting for the organizers to find someone who had access to the control room, Benjamin explained that the hum from the air conditioning system would interfere with his calculations. "I recite the numbers in my head to store them during the calculation," he said. "I have to be able to hear them, otherwise I forget them. Certain noises get in the way." In other words, one of Benjamin's secrets as a human calculator was his highly efficient use of linguistic patterns—the sounds of the numbers as they echoed in his mind.

Although few of us can match Benjamin's ability to calculate square roots of six-digit numbers, we all use the human ability to remember a spoken linguistic pattern when we learn our multiplication table. We learn by reciting the table over and over. Even today, forty-five years after I "learned my tables," I still recall the product of any two single-digit numbers by reciting that part of the table in my head. I remember the sound of the number words spoken, not the numbers themselves. Indeed, I believe the pattern I hear in my head is *precisely* the one I learned when I was seven years old!

Despite many hours of practice, most people encounter great difficulty with the multiplication tables. Ordinary adults of average intelligence make mistakes roughly 10 percent of the time. Some multiplications, such as 8×7 or 9×7, can take up to two seconds, and the error rate goes up to 25 percent. (The answers are $8 \times 7 = 54$ and $9 \times 7 = 64$. Or are they? Oh dear! I'll leave it to you to sort out.)

Why do we have such difficulty? Discounting the one times table and the ten times table as presenting no difficulty, the entire collection of multiplication tables amounts to only sixty-four separate facts (each one of 2, 3, 4, . . . , 9 multiplied by each one of 2, 3, 4, . . . , 9). Most people have little problem with the two times table or the five times table. Discounting their entries leaves just thirty-six single-digit multiplications where it takes some effort to commit them to memory. (Each of 3, 4, 6, 7, 8, 9 times each of 3, 4, 6, 7, 8, 9.) In fact, anyone who remembers that you can swap the order in multiplication (for example, 4×7 is the same as 7×4) can cut the total in half, to eighteen. So, the total number of individual facts that have to be learned to master all the multiplication tables is eighteen. That's all, eighteen!

To put that figure of eighteen simple facts in perspective, consider that by the age of six, a typical American child will have learned to use and recognize between 13,000 and 15,000 words. An American adult has a comprehension vocabulary of around 100,000 words and makes active and fluent use of 10,000 to 15,000. Then there are all the other things we remember: people's names, phone numbers, addresses, book titles, movie titles, etc. Moreover, we learn these facts with hardly any difficulty. We certainly don't have to recite words and their meanings over and over the way we do our multiplication tables. In short, most of the time there is nothing wrong with our memories. Except when it comes to the eighteen key facts of multiplication. Why?

Self-appointed critics of American youth seem to gain satisfaction from blaming the laziness of students. It seems clear to me, however, that there must be something about the multiplication tables that makes them all but unlearnable. Such a widespread problem with multiplication surely indicates some feature of the human brain that requires investigation, not criticism. There is something significant for us to learn here, and it's not our multiplication tables.

The human mind is a pattern recognizer. Human memory works by association—one thought leads to another. Someone mentions Grandfather, and that brings to mind the time he took us to the ball game, where we saw a woman who looked just like Aunt Alice, who went to live in Australia, which is where you find kangaroos, which reminds me of that kangaroo we saw in Bristol Zoo, and on and on it can go. The ability to see patterns and similarities is one of the greatest strengths of the human mind.

The human mind works very differently from a digital computer. Furthermore, each is highly unsuited to perform certain tasks that the other does with ease. Computers are good at precise storage and retrieval of information and exact calculation. A modern computer can perform billions of multiplications in a single second, getting each one right. But, despite an enormous investment in money, talent, and time over fifty years, attempts to develop computers that can recognize faces or indeed make much sense at all of a visual scene have largely failed. Humans, on the other hand, recognize faces and scenes with ease, because human memory works by pattern association. For the same reason, however, we can't do some things that computers do with ease, including remembering multiplication tables.

The reason we have such trouble is that we remember the tables linguistically, and as a result many of the different entries interfere with one another. A computer is so dumb that it sees $7 \times 8 = 56$, $6 \times 9 = 54$, and $8 \times 8 = 64$ as quite separate and distinct from each other. But the human mind sees similarities between these three multiplications, particularly linguistic similarities in the rhythm the words make when we recite them out loud. Our difficulty in trying to keep these three equations separate does not indicate a weakness of our memory but one of its major strengths—its ability to see similarities. When we see the pattern 7×8, it activates several patterns, among which are likely to be 48, 56, 54, 45, and 64.

Dehaene makes this point brilliantly in *The Number Sense* with the following example: suppose you had to remember the following three names and addresses:

- Charlie David lives on Albert Bruno Avenue.
- Charlie George lives on Bruno Albert Avenue.
- George Ernie lives on Charlie Ernie Avenue.

Remembering just these three facts looks like quite a challenge. There are too many similarities, and as a result each entry interferes with the others. But these are just entries from the multiplication tables. Let the names Albert, Bruno, Charlie, David, Ernie, Fred, and George stand for the digits 1, 2, 3, 4, 5, 6, and 7, respectively, and replace the phrase "lives on" by the equals sign, and you get the three multiplications:

- $3 \times 4 = 12$
- $3 \times 7 = 21$
- $7 \times 5 = 35$

It's the pattern interference that causes our problems.

Pattern interference is also the reason why it takes longer to realize that $2 \times 3 = 5$ is false than to realize that $2 \times 3 = 7$ is wrong. The former equation is correct for addition ($2 + 3 = 5$), and so the pattern "2 and 3 makes 5" is familiar to us. There is no familiar pattern of the form "2 and 3 makes 7."

We see this kind of pattern interference in the learning process of young children. By the age of seven, most children know by heart many additions of two digits. But as they start to learn their multiplication tables, the time it takes them to answer a single-digit addition sum increases, and they start to make errors such as $2 + 3 = 6$.

Linguistic pattern similarities also interfere with retrieval from the multiplication table when we are asked for 5×6 and answer 36 or 56. Somehow, reading the 5 and the 6 brings to mind both incorrect answers. People do not make errors such as $2 \times 3 = 23$ or $3 \times 7 = 37$. Because the numbers 23 and 37 do not appear in *any* multiplication table, our associative memory does not bring them up in the context of multiplication. But 36 and 56 are both in the table, so when our brain sees 5×6, both are activated.

In other words, much of our difficulty with multiplication comes from two of the most powerful and useful features of the human mind: pattern recognition and associative memory.

To put it another way, millions of years of evolution have equipped us with a brain that has particular survival skills. Part of that endowment is that our minds are very good at recognizing patterns, seeing connections, and making rapid judgments and inferences. All of these modes of thinking are essentially "fuzzy." Although the term "fuzzy thinking" is often used pejoratively, to mean sloppy and inadequate thinking, that is not my intended meaning here. Rather, I am referring to our ability to make sensible decisions rapidly from relatively little information. This is a powerful ability well beyond our biggest and fastest computers. Our brains are not at all suited to the kinds of precise manipulations of information that arise in arithmetic—they did not evolve to do arithmetic. To do arithmetic, we have to marshal mental circuits that developed (i.e., were selected for during our evolution) for quite different reasons. It's like using a small coin to drive in a screw. Sure, you can do it, but it's slow and the outcome is not always perfect.

We learn the multiplication tables by using our ability to remember patterns of sound. So great is the effort required to learn the tables (because of the interference effects) that people who learn a second language generally continue to do arithmetic in their first language. No matter how fluent they become in their second language—and many people reach the stage of thinking entirely in whichever language they are conversing in—it's easier to slip back into their first language to calculate and then translate the result back, than to try to relearn the multiplication table in their second language. This formed the basis of an ingenious experiment Dehaene and his colleagues performed in 1999 to confirm that we use our language faculty to do arithmetic.

The hypothesis they set out to establish was this: that arithmetical tasks that require an exact answer depend on our linguistic faculty—in particular, they use the verbal representations of numbers—whereas tasks that involve estimation or require an approximate answer do not make use of the language faculty.

To test this hypothesis, the researchers assembled a group of English-Russian bilinguals and taught them some new two-digit addition facts in one of the two languages. The subjects were then tested in one of the two languages. For questions that required an exact answer, when both the instruction and the question were in the same language, subjects answered in 2.5 to 4.5 seconds, but took a full second longer when the languages were different. The experimenters conclude that the subjects used the extra second to translate the question into the language in which the facts had been learned.

When the question asked for an approximate answer, however, the language of questioning did not affect the response time.

The researchers also monitored the subjects' brain activity throughout the testing process. When the subjects were answering questions that asked for approximate answers, the greatest brain activity was in the two parietal lobes—the regions that house the number sense and support spatial reasoning. Questions requiring an exact answer, however, elicited far more activity in the frontal lobe, where speech is controlled.

Altogether, the result was pretty convincing. The ability of humans to extend the intuitive number sense (which is not unique to humans) to a capacity to perform exact arithmetic (which does appear to be uniquely human) seems to depend on our language faculty. But if that is true, wouldn't we expect to see differences in arithmetical ability from one country to another? If the words used for numbers are significantly different, shouldn't this affect how well people learn their tables?

This is indeed what happens.

THE CHINESE ADVANTAGE

Every few years, the newspapers report that American schoolchildren have scored poorly in yet another international comparison of mathematical ability. Although there is never any shortage of knee-jerk reactions to such news, it is in fact extremely difficult to draw reliable conclusions from cross-national and cross-cultural comparisons. Many factors are involved, and even if there is a real problem, simplistic solutions are unlikely to have much effect. Education and learning are not simple matters, nor is the relationship between the two.

Chinese and Japanese children consistently outperform American children on these tests. They also outperform children from much of Western Europe, who tend to do about as well as the Americans. Given the cultural similarities between the United States and Western Europe, and the difference between those cultures and the ones of China and Japan, it is reasonable to suppose that cultural differences contribute to the disparity. Differences among the school systems are surely involved as well. But so too is language. Doing arithmetic, and in particular learning multiplication tables, is simply easier for Chinese and Japanese children, because their number words are much shorter and simpler—generally a single, short syllable such as the Chinese *si* for 4 and *qi* for 7.

The grammatical rules for building up number words in Chinese and Japanese are also much easier than in English or other European languages. For instance, the Chinese rule for making words for numbers past ten is simple: 11 is *ten one,* 12 is *ten two,* 13 is *ten three,* and so on, up to *two ten* for 20, *two ten one* for 21, *two ten two* for 22, etc. Think how much more complicated is the English system. (It's even worse in French and German, with their *quatre-vingt-dix-sept* for 97 and *vierundfünfzig* for 54.) A recent study by Kevin Miller showed that language differences cause English-speaking children to lag a whole year behind their Chinese counterparts in learning to count. By the age of four, Chinese children can generally count up to 40. American children of the same age can barely get to 15, and it takes them another year to get to 40. How do we know the difference is due to language? Simple. The children in the two countries show no age difference in their ability to count from 1 to 12. Differences appear only when the American children start to encounter the various special rules for forming number words. The Chinese children, meanwhile, simply keep applying the same ones that worked for 1 to 12. (American children often apply the same rules, but they find they have made a mistake when they try to use words like twenty-ten and twenty-eleven.)

In addition to being easier to learn, the Chinese number word system also makes elementary arithmetic easier, because the language rules closely follow the base-10 structure of the Arabic system. A Chinese pupil can see from the linguistic structure that the number "two ten five" (i.e., 25) consists of two 10s and one 5. An American pupil has to *remember* that "twenty" represents two 10s, and hence that "twenty-five" represents two 10s and one 5.

It is remarkable how often we dragoon our linguistic ability into helping us learn important number facts (such as multiplication tables) and do arithmetic. But as often happens when we take a tool developed for one purpose and use it for another, the outcome can be less than ideal.

WHY MATHEMATICS SEEMS ILLOGICAL

To be able to use numbers properly, it is not enough to know how to manipulate the symbols according to the rules. You also have to relate the symbols and your manipulations of them to your innate sense of number—to the numerical quantities the symbols denote. Otherwise, because our minds automatically see patterns, we can easily find ourselves performing nonsensical manipulations of symbols.

For example, the following incorrect sum illustrates a common error in adding fractions:

$$1/2 + 3/5 = 4/7$$

A person who makes this mistake sees this as two addition sums, first adding the numerators $1 + 3 = 4$ and then adding the denominators $2 + 5 = 7$. Symbolically, this is the most logical thing to do. It is incorrect because it makes no sense in terms of the numbers the symbols represent. The symbolic manipulations you have to perform to get the correct answer—the ones that correspond to adding the actual fractional numbers represented by the symbol words $1/2$ and $3/5$—are fairly complicated. What is more, those symbol-manipulation rules *only* make sense if you think of the numbers the symbols represent. Purely as rules for manipulating symbols, they make no sense at all.

It is undoubtedly because of such cases that many children come to see mathematics as "illogical" and "full of rules that make no sense." They think of mathematics as a collection of rules for *doing things with symbols.* Some of these rules make sense; others seem quite arbitrary. The only way to avoid this misconception is for teachers to ensure that their pupils understand what the symbols represent. This is rarely done. How then do some children learn how to add fractions correctly?

Since the human mind is an excellent pattern recognizer with tremendous adaptive powers, with enough training it can learn to perform almost any symbolic procedure in an essentially "mindless" fashion. Thus, it is possible to memorize a procedure for manipulating the symbols to add fractions correctly:

> Start out by multiplying the two denominators. That will give you the denominator in the answer. Then multiply the numerator of the first fraction by the denominator of the second, and the numerator of the second fraction by the denominator of the first, and add those two results. That gives you the numerator in the answer. Then see if there are any numbers that divide both the numerator and the denominator in your answer, and if there are, divide both numerator and denominator by that number. Repeat this double division until you can't find any such common divisors. What's left is your final answer.

This procedure looks complicated—and at a symbolic (i.e., linguistic) level it makes no sense at all. But with practice most people can learn to follow it. Evolution has equipped us with a brain that is actually pretty good at learning particular sequences of actions. But unless someone, at each step, shows you what is going on in terms of the *numbers* represented by the symbols, the whole thing is just so much mumbo-jumbo. Learn how to perform the mumbo-jumbo and you get an A. How many children leave school with good grades in mathematics but no understanding of what they were doing? Surely a lot, judging from the large numbers of perfectly intelligent adults who cannot add fractions. If only they understood what was going on, they would never forget how to do it. Without such understanding, however, few can remember such a complicated procedure for long once the final exam has ended.

And without understanding, there is little mystery why, for many people, elementary school arithmetic is well summed up by Lewis Carroll's satirical terms *ambition*, *distraction*, *uglification*, and *derision*.

Here are some other difficulties that arise from blindly applying a symbolic rule without linking the symbols to the numbers they represent. Answer the following questions:

- A farmer has 12 cows. All but 5 die. How many cows remain?
- Tony has 5 balls, which is 3 fewer than Sally. How many balls does Sally have?

Readers of this book might well score above average on these. But I know from experience that many intelligent people get one or both of them wrong. The numbers 12 and 5 in the first problem, together with the question "How many remain?" create a strong temptation to perform the subtraction $12 - 5 = 7$ and give 7 as the answer. The correct answer is 5, but to get it you have to think what the problem is actually saying. Blindly rushing to the symbolic manipulation stage sometimes works, but not here. Overall it is a disastrous strategy to follow. The symbols are there to *assist* our reasoning, not eliminate it.

Similarly for the second problem. You see the numbers 5 and 3 together with the word "fewer" and the temptation is to perform the subtraction $5 - 3 = 2$. Again, a hasty leap to symbolic manipulation has led to the wrong answer. When you stop and think what the question is saying, you realize you should *add* 3 to 5, giving the correct answer that Sally has $5 + 3 = 8$ balls.

People who are "good at arithmetic" do not make such mistakes. What sets them apart from the many people who never seem to "get it" is not that they have memorized the rules better. Rather, they *understand* those rules. Indeed, their understanding is such that they don't really have need for the rules at all. This is true for the many people for whom numbers bring no fear, but the most dramatic illustration is provided by the lightning calculators, such as Arthur Benjamin, whom we met briefly earlier, who are able to perform highly complicated calculations rapidly in their heads.

Part of the calculating wizards' secret is that, for them, many numbers have meaning. For most of us who are comfortable with numbers, a number such as 587 doesn't *mean* anything—it's just a number. But to a calculating wizard, that number word 587 may well have meaning—it may conjure up a mental image—just as the English word "cat" has meaning for us and conjures up an image in our minds.

Some numbers, of course, do have meaning for us. Americans see meaning in the numbers 1492 (Columbus's discovery of America) and 1776 (the signing of the Declaration of Independence); Britons see meaning in 1066 (the Battle of Hastings); and anyone with a technical education sees meaning in the number 314159 (the start of the decimal representation of the mathematical constant π). Other numbers that have meaning for us—and which we therefore remember—are our birthdate and our telephone number.

For a calculating wizard, however, many numbers have meaning. Generally, that meaning lies not in the everyday world of dates, ID cards, and telephone numbers, but in the world of mathematics itself. For instance, Wim Klein, a famous calculating wizard who in the days before electronic computers once held a professional position with the title "computer," observes, "Numbers are friends to me." Referring to the number 3844, he says, "For you it's just a three and an eight and a four and a four. But I say, 'Hi, 62 squared!'"

Because numbers have meaning to Klein and the other calculating wizards, calculation is meaningful to them. Consequently, for reasons I shall elaborate in due course, they are much better at it.

Of course, lightning calculators are a very special case. Most of us would not want to develop such a skill even if we could. But the point is that arithmetic becomes much easier when you understand what it is about—when the symbols mean something to you. Not seeing the meaning is the main reason so many people say that they are "no good at math."

I end this chapter with two notes of caution.

First caution: I have tried to show how our facility for language—particularly our ability to recognize and remember linguistic patterns—helps us do mathematics, particularly basic arithmetic. I am *not* saying that we do mathematics using ordinary language. Rather, the examples I have given show how we sometimes use linguistic skills as a tool to *assist* us.

In fact, based on subjective reports by mathematicians, myself included, it is almost certainly the case that we do not use language to actually *do* mathematics. We do use language to record and convey the results of our thinking—indeed, on occasion to convey our actual thought processes. But the thinking process itself, what we generally call "doing mathematics," is not linguistic.

Second caution: I announced at the very beginning of this book that my goal is to convince you that the ability to do mathematics is based on our facility for language. The discussions in this chapter have virtually nothing to do with that overall claim. It will in fact take me some considerable time to make my case—that's why this is a book and not a short pamphlet. For now, let me reiterate that the principal claim in this book is not that we use language to do mathematics, but that the feature of our brain that enables us to use language is the same feature that makes it possible for us to do mathematics. Thus when the human brain developed the ability to use language, it automatically acquired the ability to do mathematics. When I come to spell out my case in detail, it will be at a fairly deep level, involving the *nature* of mathematics and the *nature* of language. It's time to take a look at the first of these: the nature of mathematics. Not number, *mathematics*.

· 4 ·

WHAT IS THIS THING CALLED MATHEMATICS?

TO MOST PEOPLE, mathematics is calculating with numbers. By concentrating in the previous two chapters on numerical ability—and the lack of it—I may even have unintentionally reinforced this myth. I shall put that right at once and dispose of a number of other myths while I'm at it. For example:

- Mathematicians have a good head for figures. (Some do, some don't.)
- Mathematicians like adding up long columns of numbers in their head. (Surely, no one likes this.)
- Mathematicians find it easy to balance their checkbook. (I don't, for one.)
- Mathematicians revel in solving ten simultaneous linear equations in ten unknowns. In their heads. (I actually did enjoy such stunts as a high school student, but I grew out of it.)

- College-level mathematics students all become mathematics teachers or accountants when they graduate. (Some do, but many do something else.)
- Mathematicians are not creative. (If you believe this, you certainly don't know what mathematics is about.)
- There is no such thing as beauty in mathematics. (Philistine!)
- Mathematics is predictable. It involves following precise rules. (Like music, drama, sculpture, painting, writing novels, chess, and football?)
- In mathematics, there is always a right answer. (And it's in the back of the book.)

I shall correct these myths not by attacking them in turn (other than the parenthetical comments I appended to each one), but by providing a glimpse of what mathematics is really about. To anyone who understands what mathematics is, each of the assertions listed above is so obviously incorrect that no refutation is necessary. If you find yourself shaking your head at this, then all I ask is that, for the rest of this chapter, you agree to suspend whatever impression you have of what mathematics is.

THE NATURE OF THE BEAST

The best one-line definition of mathematics that I know is the one I gave in chapter 1: *mathematics is the science of patterns.*

The phrase is not mine. I first saw it in print as the title of an article in *Science* magazine, written by the mathematician Lynn Steen in 1988. Steen admits it did not originate with him. The earliest written source I have found is W. W. Sawyer's 1955 book *Prelude to Mathematics:*

> For the purpose of this book we may say, "Mathematics is the classification and study of all possible patterns." Pattern is here used in a

way that not everybody may agree with. It is to be understood in a very wide sense, to cover almost *any kind of regularity that can be recognized by the mind.* Life, and certainly intellectual life, is only possible because there are certain regularities in the world. A bird recognizes the black and yellow bands of a wasp; man recognizes that the growth of a plant follows the sowing of seed. In each case, a mind is aware of pattern. (p. 12; emphasis in the original)

I first read *Prelude to Mathematics* as a science-mad student in high school. Sawyer's depiction of what mathematics is really about—which differed greatly from the impression I had formed from my school lessons—so captivated and intrigued me that for the first time I contemplated becoming a mathematician myself.

But I was not aware of having read that passage about patterns until 1995. In 1994, I wrote a Scientific American Library book with the title *Mathematics: The Science of Patterns.* The following year, a reader wrote to me and pointed out the similarity between my title and Sawyer's words. Obviously, his book had influenced me, with the result that, years later when I saw Steen's article, the phrase "science of patterns" resonated with me at once.

Andrew Gleason of Harvard University has put forward a similar view of mathematics. In an article published in the *Bulletin of the American Academy of Arts and Sciences* in October 1984, he wrote:

Mathematics is the science of order. Here, I mean order in the sense of pattern and regularity. It is the goal of mathematics to identify and describe sources of order, kinds of order, and the relations between the various kinds of order that occur.

Much of the impact of the phrase "the science of patterns" comes from its brevity. But brevity comes at a price of possible misunderstanding. In this case, the word "patterns" requires some elaboration. It certainly is not restricted to visual patterns such as wallpaper patterns or the pattern of tiles on a bathroom floor, although both can be studied mathematically. A slightly fuller definition would be: *mathematics is the science of order, patterns, structure, and logical relationships*. But since the meaning mathematicians attach to the word "pattern" in this context includes all the terms in the expanded definition, the shorter version says it all—provided you understand what is meant by "pattern."

The patterns and relationships studied by mathematicians occur everywhere in nature: the symmetrical patterns of flowers, the often complicated patterns of knots, the orbits swept out by planets as they move through the heavens, the patterns of spots on a leopard's skin, the voting pattern of a population, the pattern produced by the random outcomes in a game of dice or roulette, the relationship between the words that make up a sentence, the patterns of sound that we recognize as music. Sometimes the patterns are numerical and can be described using arithmetic—voting patterns, for example. But often they are not numerical—for example, patterns of knots and symmetry patterns of flowers have little to do with numbers.

Because it studies such abstract patterns, mathematics often allows us to see—and hence perhaps make use of—similarities between two phenomena that at first appear quite different. Thus, we can think of mathematics as a pair of conceptual spectacles that enable us to see what would otherwise be invisible—a mental equivalent of the physician's X-ray machine or the soldier's night-vision goggles. With mathematics, we can *make the invisible visible*—another phrase that I found so powerful that I took it as the subtitle of another book I wrote: *The Language of Mathematics: Making the Invisible Visible*. Let me give some examples of how mathematics makes the invisible visible.

Without mathematics, there is no way you can understand what keeps a jumbo jet in the air. As we all know, large metal objects don't remain off the ground without something to support them. But when you look at a jet aircraft flying overhead, you can't see anything holding it up. It takes mathematics—in this case an equation discovered by the mathematician Daniel Bernoulli early in the eighteenth century—to "see" what keeps an airplane aloft.

What is it that causes objects other than aircraft to fall to the ground when we release them? "Gravity," you answer. But that's just giving it a name. It's still invisible. We might as well call it magic. Newton's equations of motion and mechanics in the seventeenth century enabled us to "see" the invisible forces that keep the earth rotating around the sun and cause an apple to fall from the tree onto the ground.

Both Bernoulli's equation and Newton's equations use calculus. Calculus works by making visible the infinitesimally small. That's another example of making the invisible visible.

Here's another. Two thousand years before we could send spacecraft into outer space to photograph our planet, the Greek mathematician Eratosthenes used mathematics to show that the earth was round. Indeed, he calculated its diameter, and hence its curvature, with considerable accuracy.

Physicists are using mathematics to try to see the eventual fate of the universe. In this case, the invisible that mathematics makes visible is the invisible of the not-yet-happened. They have already used mathematics to see into the distant past, making visible the otherwise invisible moments when the universe was first created in what we call the Big Bang.

Coming back to earth at the present time, how do you "see" what makes pictures and sound of a football game miraculously appear on a television screen on the other side of town? One answer is that the pictures and sound are transmitted by radio waves—a kind of electromagnetic radiation. But as with gravity, that just gives the phenomenon a name, it doesn't help us to "see" it. You need Maxwell's equations, discovered in the last century, to "see" the otherwise invisible radio waves.

Here are some human patterns:

- Aristotle used mathematics to try to "see" the invisible patterns of sound that we recognize as music.

- Aristotle also used mathematics to try to describe the invisible structure of a dramatic performance.
- In the 1950s, the linguist Noam Chomsky used mathematics to "see" the invisible, abstract patterns of words that we recognize as a grammatical sentence. He thereby turned linguistics from an obscure branch of anthropology into a thriving mathematical science.

Finally, using mathematics, we are able to look into the future:

- Probability theory and mathematical statistics let us predict the outcomes of elections, often with remarkable accuracy.
- We use calculus to predict tomorrow's weather.
- Market analysts use mathematical theories to predict the behavior of the stock market.
- Insurance companies use statistics and probability theory to predict the likelihood of an accident during the coming year, and set their premiums accordingly.

When it comes to looking into the future, our mathematical vision is not perfect. Our predictions are sometimes wrong. But without mathematics, we cannot even see poorly.

Just under forty years ago, the scientist Eugene Wigner wrote an article titled "The Unreasonable Effectiveness of Mathematics in the Natural Sciences." "Why," asked Wigner, "is it the case that mathematics can so often be applied, and to such great effect?" Once you realize that mathematics is not some game that people make up, but is about the patterns that arise in the world around us, Wigner's observation does not seem so surprising.

Mathematics is not about numbers, but about life. It is about the world in which we live. It is about ideas. And far from being dull and sterile, as it is so often portrayed, it is full of creativity.

Many people have compared mathematics to music. Certainly the two have a lot in common, including the use of an abstract notation. Indeed, the first thing that strikes anyone who opens a typical book of mathematics is that it is full of symbols—page after page of what looks like a foreign language written in a strange alphabet. In fact, that's exactly what it is. Mathematicians express their ideas in the language of mathematics.

Why? If mathematics is about life and the world we live in, why do mathematicians use a language that turns many people off the subject before they are out of high school? It's not because mathematicians are perverse individuals who like to spend their days swimming in an algebraic sea of meaningless symbols. The reason for the reliance on abstract symbols is that the patterns studied by the mathematician are abstract patterns.

You can think of the mathematician's abstract patterns as "skeletons" of things in the world. The mathematician takes some aspect of the world, say a flower or a game of poker, picks some particular feature of it, and then discards all the particulars, leaving just an abstract skeleton. In the case of the flower, that abstract skeleton might be its symmetry. For a poker game, it might be the distribution of the cards or the pattern of betting.

To study such abstract patterns, the mathematician has to use an abstract notation. Musicians use an equally abstract notation to describe the patterns of music. Why do they do this? Because they are trying to describe on paper a pattern that exists only in the human mind. The same tune can be played on a piano, a guitar, an oboe, or a flute. Each produces a different sound, but the tune is the same. What distinguishes the tune is not the instrument but the *pattern of notes* produced. It is that abstract pattern that is captured by musical notation.

When a mathematician looks at a page of mathematical symbols, she does not "see" the symbols, any more than a trained musician "sees" the musical notes on a sheet of music. The trained musician's eyes read straight through the musical symbols to the sounds they represent. Similarly, a trained mathematician reads straight through the mathematical symbols to the patterns they represent.

In fact, the connection between mathematics and music may go deeper—right to the structure of the very device that creates both: the human brain. Using modern imaging techniques that show which parts of the brain are active while the subject performs various mental or physical tasks, researchers have compared the brain images produced by professional musicians listening to music with those of professional mathematicians working on a mathematical problem. The two images are very similar, showing that expert musicians and expert mathematicians appear to be using the same circuits. (The result does not always hold for amateurs.)

With that general overview behind us, let's take a closer look at some ways in which mathematics allows us to make the invisible visible. I will start with phenomena that are obviously "patterns" and then move on to increasingly sophisticated kinds of order.

HOW DO MATHEMATICIANS GET INTO SHAPE?

The word *geometry* comes from the Greek *geo metros*, meaning "earth measurement." Geometry was first developed by the ancient Egyptians, Babylonians, and Chinese to determine land boundaries, construct buildings, and plot the stars for navigation. The geometry most people over the age of thirty remember from their high school days is a refined version of the subject developed largely by the ancient Greeks between 650 and 250 BC.[1]

1 Younger people may not have taken a geometry class. The subject was reclassified as optional some years ago in the mistaken belief that it was no longer sufficiently relevant to today's world, a view that demonstrates the ignorance of many of the people who make such decisions. Although it is true that hardly anyone ever makes direct use of geometrical knowledge, it was the only class in the high school curriculum that exposed children to the important concept of formal reasoning and mathematical proof.

Exposure to formal mathematical thinking is important for at least two reasons. First, a citizen in today's mathematically based world should have at least a general sense of one of the major contributors to society. Second, a survey carried out by the U.S. Department of Education in 1997 (the Riley Report) showed that students who completed high school geometry performed markedly better in gaining entry to college and did better when at college than those students who had not taken such a course, *regardless of the subjects studied at college*. As the survey organizers pointed out, the major factor was not how well the students do in such a course. Merely completing it gives them a tremendous advantage in all their other courses.

In geometry we study patterns of shape. Not any kind of shape, but regular shapes, such as triangles, squares, rectangles, parallelograms, pentagons, hexagons, circles, ellipses, and, in three dimensions, tetrahedra, cubes, octahedra, spheres, ellipsoids, and the like. We see examples of these shapes in the world around us—the circular appearance of the sun and the moon, circular clock faces, circular wheels, triangular, square, and hexagonal floor tiles, cubic boxes, spherical tennis balls, ellipsoidal footballs, and so forth. Geometry studies these shapes in the abstract, removed from any particular real-world example.

Having decided what kinds of shape to study, the geometrician then discovers general facts that apply to all instances of the shape. For example, Pythagoras's theorem says that for any right triangle, the square of the hypotenuse is equal to the sum of the squares of the other two sides.

Many of the basic theorems of geometry were collected by the ancient Greek mathematician Euclid in his famous book *Elements,* written around 350 BC. One of the deepest theorems in *Elements* concerns the so-called regular polyhedra. These are the three-dimensional analogs of the regular polygons.

A regular polygon is a figure made up of equal straight-line edges, each adjacent pair of which meets at exactly the same angle. The simplest such figure is an equilateral triangle, where the sides are all equal and the angle of each vertex is 60°. Then comes a square; followed by a regular pentagon (a 108° angle between touching edges); a regular hexagon (angle 120°), etc. (see Figure 4.1). A regular polygon may have any number of sides you choose.

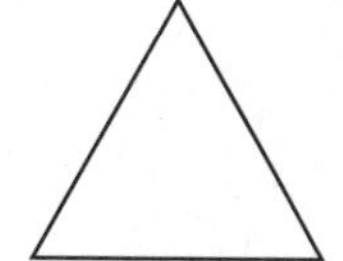
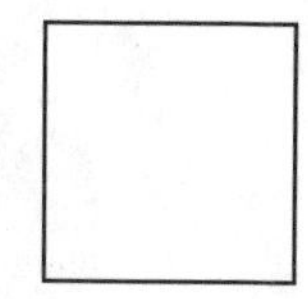
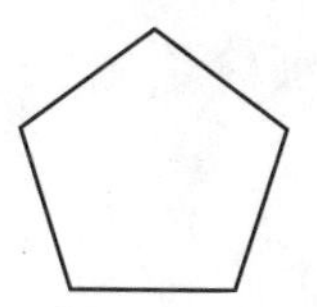
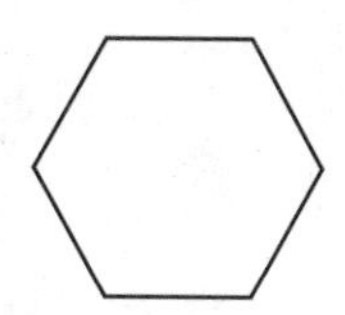

FIGURE 4.1 Regular Polygons. A polygon is regular if all its sides are equal and all its interior angles are equal. There are regular polygons of any number of sides greater than two. The first four are shown: the equilateral triangle, the square, the regular pentagon, and the regular hexagon.

A regular polyhedron is a three-dimensional object whose faces are all identical regular polygons, with the angles between all touching faces the same. You might think that, as with the regular polygons, there are regular polyhedra with any number of faces. But this is not true. As was proved in Euclid's *Elements*, there are just five regular polyhedra: the regular tetrahedron (four identical equilateral-triangle faces); the cube (six identical square faces); the regular octahedron (eight identical equilateral-triangle faces); the regular dodecahedron (twelve identical regular pentagon faces); and the isocahedron (twenty identical equilateral-triangle faces). These are all illustrated in Figure 4.2. In this instance, the laws of geometry restrict the number of possibilities.

A similar thing happens with wallpaper patterns. The laws of geometry restrict the number of possibilities to seventeen.

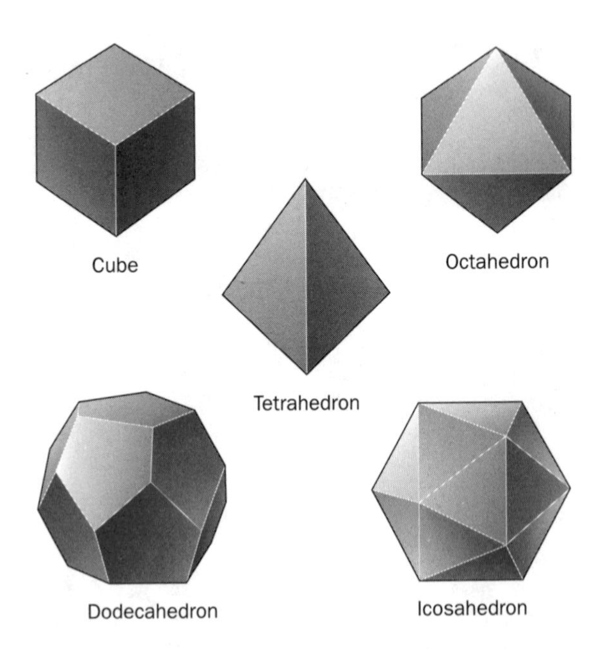

FIGURE 4.2 Regular Solids. A polyhedron is regular if all its faces are identical and all its interior angles are equal. There are exactly five regular polyhedra.

Now, whereas most of us are surprised to learn that there are only five regular polyhedra, we are not surprised to find that there is a mathematical theorem about them. Regular polyhedra are just the kind of thing you would expect to find a mathematical theorem about. But wallpaper?

It's a bit more surprising to learn of a theoretical limit to wallpaper patterns. Surely designers can always come up with new ideas, can't they? Indeed they can. But the mathematical theorem is not about the fine details of wallpaper patterns. The feature that attracts mathematicians is that wallpaper takes a particular pattern and *repeats* it without limit. How many different ways are there for a fixed pattern to repeat endlessly? What are the patterns of repetition?

Well, you can simply copy the basic pattern side by side. Or you can move it along, alternately reflecting it left to right. Or maybe you can move it with some kind of rotation? Or maybe...Well, just how many different ways *can* you do it?

Once you start to think along these lines, asking yourself what exactly is involved in endlessly repeating a fixed, basic pattern, you realize that this too is highly structured, indeed *geometric*. Just the kind of thing to which mathematics might be applied.

Even so, I still find it surprising that there are exactly seventeen ways to repeat a fixed pattern. Some of those seventeen ways are fairly intricate, but interestingly, designers of rugs, mosaics, and decorative wall tilings had discovered all of them hundreds of years before the mathematicians of the nineteenth century enumerated them all and proved that there were no others. Figure 4.3 gives examples of each of the seventeen possible wallpaper patterns.

The wallpaper-pattern theorem is a good illustration of how mathematics approaches the world. Mathematics is a precise subject that deals with precise patterns. In the case of geometry, the patterns are immediately recognizable—the straight lines, the triangles, the circles, the tetrahedra, the spheres, and so forth. But sometimes you have to find the right way to look at a phenomenon before you discover those precise patterns. In the case of

FIGURE 4.3 Wallpaper patterns. There are exactly seventeen different ways to repeat a fixed pattern indefinitely to cover the whole plane. The illustrations show one wallpaper pattern that repeats in each of the seventeen different ways.

wallpaper design, the right question to ask is: what are the patterns by which a fixed design can repeat endlessly, to cover an entire wall?

Of course, until you investigate, you can't know whether you will discover any interesting (or useful) mathematics. The repeating patterns of wallpaper are *mathematical* because it is possible to prove a theorem about them. A similar thing happened with the coat patterns of animals.

Although the spots on a leopard or the stripes on a tiger have some regularity, they don't appear to exhibit the kind of *geometrical* regularity to which mathematics can be applied. But is there a hidden pattern to animal coat markings to which we can apply mathematics?

Until recently, the answer would have been no. But during the last twenty years, mathematicians have started to develop a "geometry" of living things. One of the results of that new mathematics is the discovery that the coat patterns of animals are every bit as constrained by mathematical laws as are regular polyhedra or wallpaper patterns. The key was to discover the right way to look at the phenomenon. In the case of animal coat patterns, the question to ask is: what is the mechanism that produces the different patterns of spots and stripes?

THE GEOMETRY OF ANIMAL COAT PATTERNS

The idea of applying mathematics to study the form of living things was put forward by the great British thinker D'Arcy Thompson in his book *On Growth and Form*, first published in 1917. In the 1950s, the English mathematician Alan Turing took up Thompson's suggestion, proposing a specific mechanism for applying mathematics to study the coat patterns of animals. He called this new field "morphogenesis."

Turing's idea was to formulate equations that describe the way animal coat patterns form, based on the biological or chemical processes that generated them. It was a good idea, but for many years no one could make much progress with it. For one thing, biologists had only the beginnings of an understanding of the actual growth processes. For another, the kinds of equations that come from the relevant biology and chemistry cannot be solved in the same way that an equation for, say, a circle or a parabola can, allowing you to draw a picture (a graph) of the curve.

Only with the development of the computer—and computer graphics—was it possible to take up Turing's suggestion.

In the late 1980s, mathematician James Murray of the University of Oxford embarked on a three-step process to carry out Turing's program. The first step was to write down equations that described the chemical processes that cause coloration in an animal coat. The second step was to write a computer program to solve these equations. The third step was to use computer graphics techniques to turn those solutions into pictures. If everything worked—and if the equations really did describe the growth patterns—then the resulting pictures should resemble some (all?) of the kinds of animal coat patterns observed in nature.

Murray knew that any coloration of an animal's coat is caused by a chemical called melanin, which is produced by cells just beneath the surface of the skin. It's the same chemical that makes fair-skinned people develop a suntan. But why spots on the leopard? Or stripes on the tiger?

Murray began by assuming that certain chemicals stimulated the cells to produce melanin. The visible coat pattern was thus a reflection of an invisible chemical pattern in the skin: high concentrations of the chemicals give rise to melanin coloration, low concentrations leave the skin largely uncolored. The question then was: what caused the melanin-inducing chemicals to cluster into a regular pattern so that when they "switched on" the melanin, the result was a visible pattern in the skin?

One possible mechanism was provided by so-called reaction-diffusion systems. A reaction-diffusion system is where two or more chemicals in the same solution (or in the same skin) react and diffuse through the solution, fighting with each other for territory. Although first proposed by mathematicians as a theoretical idea in the 1950s, reaction-diffusion systems were only later observed by chemists in the laboratory. Even today, they are still studied more by mathematicians, in a theoretical way, than by chemists in the laboratory.

To keep things as simple as possible, Murray assumed that just two chemicals are produced in the skin, one of which stimulates melanin production, the other which inhibits it. He further assumed that the presence of the stimulating chemical triggers increased production of the inhibitor. Finally, he supposed that the inhibiting chemical diffused through the skin faster than the stimulator. Given these assumptions, if a concentration of the stimulating chemical is formed, triggering production of the inhibitor, then the faster-moving inhibitor would be able to encircle the more slowly diffusing stimulator and prevent further expansion. The result would be a region of stimulator kept in check by an encircling ring of inhibitor. In other words, a spot.

Murray likens this process to the following scenario. Imagine, he says, a very dry forest in which scattered fire crews are stationed with helicopters and firefighting equipment. When a fire breaks out (the stimulator), the firefighters (the inhibitors) spring into action. Traveling in their helicopters, they can move much more quickly than the fire. (The inhibitor diffuses faster than the stimulator.) However, because of the intensity of the fire (the high concentration of the stimulator), the firefighters cannot contain the fire at its core. So, using their greater speed, they outrun the front of the fire and spray fire-resistant chemicals onto the trees. When the fire reaches the sprayed trees, its progress is stopped. Seen from the air, the result will be a blackened spot where the fire burned, surrounded by the green ring of sprayed trees.

Now imagine that several fires break out all over the forest. Seen from the air, the resulting landscape will show patches of blackened, burned trees interspersed with the green of unburned trees. If the fires break out sufficiently far apart, the resulting aerial pattern could be one of black spots in a sea of green. But if nearby fires are able to merge before being contained, different patterns could result. The exact pattern will depend on the number and relative positions of the initial fires and the relative speeds of the fire and the firefighters (the reaction-diffusion rates).

The case of interest to Murray was what the resulting pattern would be if the initial fire pattern was random. How would different reaction-diffusion rates then affect the final pattern? More specifically, were there rates that, starting from a random pattern of fire sources, would lead to recognizable patterns such as spots or stripes? This is where mathematics came in. There are equations that describe how chemicals react and disperse. They involve techniques from calculus and are called partial differential equations. Given the equations, Murray was able to forget the biology and the chemistry and concentrate on mathematics. By putting his equations into a computer, Murray was able to generate pictures on the screen, showing the way the chemicals dispersed. To his surprise, even with the simple scenario of just two chemicals, his equations generated dispersal patterns that looked remarkably like the skins of animals.

In fact, by experimenting with various parameters in his equations, Murray discovered a simple, hitherto unsuspected relationship between the shape and size of the skin area on which the reaction-diffusion process occurs and the coat pattern that results. Very small skin regions, he found, led to no pattern at all; long, thin regions led to stripes perpendicular to the length of the region; and squarish regions of roughly the same overall area gave rise to spots whose exact pattern depended on the region's dimensions. For larger area he got no pattern at all. (See Figure 4.4.)

The key factor is the shape and size of the animal's coat at the time the reaction-diffusion process occurs, which for most animals is during embryonic development—not the shape and size of the fully grown adult. For example, there is a four-week period early in the year-long gestation of a zebra during which the embryo is long and pencil-like. Murray's mathematics predicts that, if the reactions take place during this period, the resulting pattern will be stripes. Leopard embryos, however, are fairly chubby when reaction-diffusion occurs, and Murray's equations predict spots. Apart, that is, from the tail. The tail is long and pencil-like throughout development, which explains why the tail of the leopard is always striped.

Do Murray's equations describe what actually happens? Since biologists have not yet observed reaction-diffusion experimentally in embryo skin, we cannot know. Murray chose the simplest case, with just two chemical reagents, so there is no reason to suppose his equations describe the situation exactly. On the other hand, those equations do produce all the coat patterns found in nature.

Some of the most convincing evidence that Murray is on the right track is that his mathematics answers a long-standing question in zoology: why is it that several kinds of animals have spotted bodies and striped tails, but none have striped bodies and spotted tails? There seems to be no evolutionary reason for this curious fact. Murray provides a ridiculously

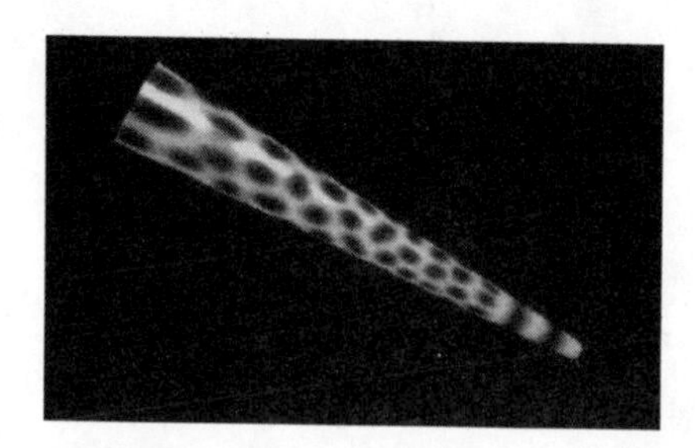

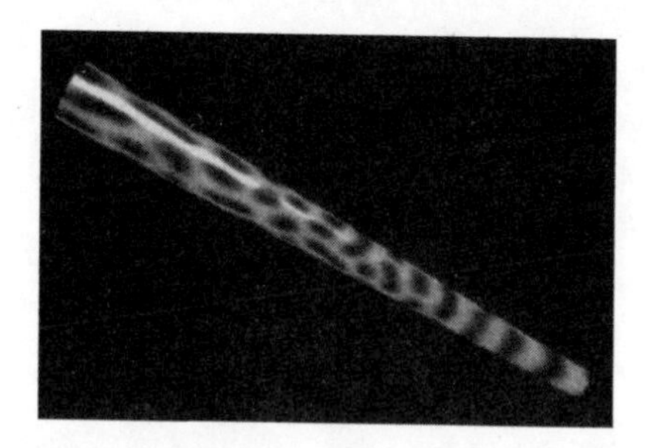

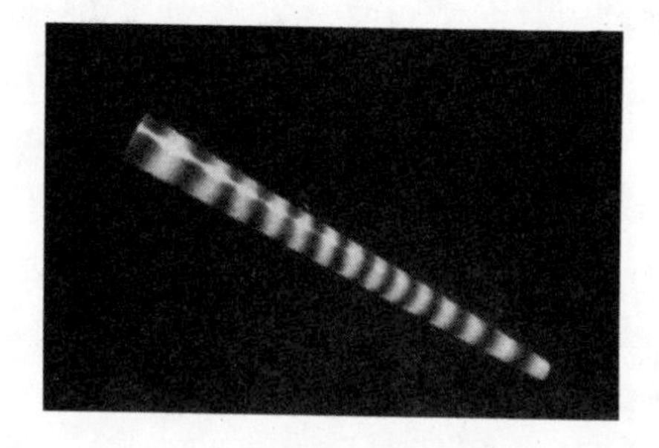

FIGURE 4.4 Animal coat patterns produced mathematically on a computer screen, by solving the equations written down by mathematical biologist James Murray. By changing the value of a single parameter in his equations, Murray could change a spotted tail into a striped one.

simple explanation: it's a direct consequence of the fact that many animal embryos have chubby bodies and skinny tails, but no animal embryo has a long skinny body and a chubby tail.

If Murray *is* correct, then we have a marvelous example of how evolution leads to a highly efficient process. The two obvious survival benefits of animal coat patterns are camouflage and appeal to the opposite sex. The question is, how is this achieved?

Instead of coding the entire detailed pattern into the animal's DNA, nature could more efficiently use the mathematical patterns Murray has discovered. All that would be necessary would be for the DNA to encode instructions telling the skin of the developing embryo (or newly born creature in the case of animals, such as Dalmatian dogs, whose skin pattern appears after birth) when to activate the reaction-diffusion process and when to bring it to a halt. The final coat pattern would then be determined by the shape and size of the skin at that stage of development.

The mathematics of animal coat patterns is just one of several new kinds of "geometry of life" described in my 1998 book *Life by the Numbers.* Another example I think would be usefully presented here is geometry of flowers.

THE GEOMETRY OF FLOWERS

How do you describe the shape of a flower? You could say a daisy is circular. But no flower is really circular. It only looks that way from a distance. When you take a proper look, you see that the flower is made up of many petals, which trace out a shape much more complicated than a circle. Can mathematics be used to describe the real shape of a daisy? Or how about a flower that is not circular, such as a lilac? Can mathematics describe the shape of a lilac flower?

The question might seem pointless. After all, what possible benefit is to be gained from giving a mathematical description of a flower?

One answer, which history has taught us again and again, is that scientific knowledge generally turns out to be beneficial. For example, in the early nineteenth century, mathematicians began to study the patterns of knots. Their only motivation was curiosity. But during the last twenty-five years, biologists have used the mathematics of knots to help in the fight against viruses, many of which alter the way a DNA molecule wraps around itself to form a knot.

A possible benefit of finding a mathematical description of a lilac is that it could lead to more accurate weather forecasts. Here's how.

As in the case of wallpaper patterns and animal coat markings, the crucial first step is to find the right way of looking at the phenomenon. In the case of animal coats, the trick was to concentrate not so much on the final markings but on the process that led to those markings. Maybe a similar approach will work for flowers. Can we develop a geometry of flowers by looking at the way nature might *create* the flower's shape?

If you look closely at a lilac, you will notice that a small part looks much the same as the entire flower. You see the same phenomenon with certain other flowers, and with some vegetables such as broccoli or cauliflower. Mathematicians refer to this property as self-similarity.

Clouds also have self-similarity. A mathematical way to describe self-similar patterns could be used to study clouds. Given a mathematical description of clouds, we could simulate the formation, growth, and movement of clouds on a computer. Using those simulations, maybe we could improve our ability to forecast severe weather, using our forecasts to protect ourselves better from the consequences of a major storm or a tornado. This is not completely fanciful. Researchers have been carrying out just such investigations for some years.

A mathematician named Helge Koch studied self-similarity at the end of the nineteenth century. Koch noticed that if you take an equilateral triangle, add a smaller equilateral triangle to the middle third of each side, then repeat the process of adding smaller and smaller triangles to the middle-thirds of the sides, eventually you will develop the fascinating shape now called the Koch snowflake, shown in Figure 4.5. (To be precise, you delete the middle-third length each time you add a new triangle.)

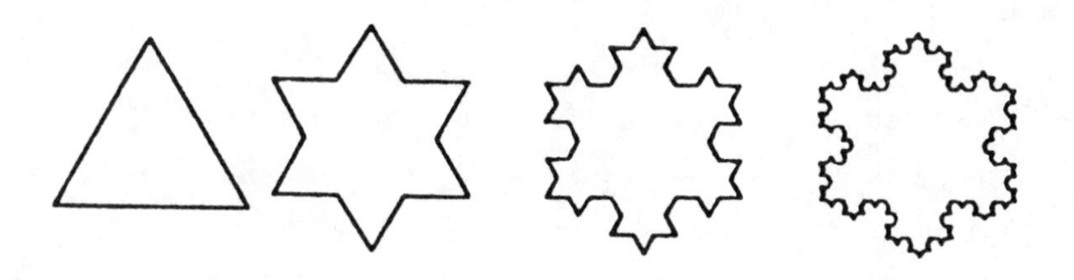

FIGURE 4.5 The Koch snowflake starts to take shape.

This example shows that a complicated-looking shape can result from the repeated application of a very simple rule. The self-similarity results from using the same rule over and over again. Present-day mathematicians refer to self-similar figures as fractals, a name invented in the 1960s by the mathematician Benoit Mandelbrot, who cataloged and studied many instances of self-similarity in nature.

To obtain a geometric description of a fractal, the mathematician looks for rules that, when used over and over again, produce its self-similar shape. Such a system of repeatable growth rules is called an L-system, after Aristid Lindenmayer, a biologist who, in 1968, developed a formal model for describing the development of plants at the cellular level.

For example, a very simple L-system to produce a tree-like shape might say that if we start with the top part of any branch, that portion forms two new branches, giving three new tops. When we repeat this rule, we find we rapidly get a tree-like shape. It is easy to carry out the first few iterations of this rule using a paper and pencil (see Figure 4.6). But you only start to get something that looks like a real tree when you apply the rule hundreds of times on a computer.

Mathematician Przemyslaw Prusinkiewicz has used such an approach to generate flowers on a computer. To create, say, a lilac, Prusinkiewicz starts with a very simple L-system to generate the skeleton of the flower. Then, by taking careful measurements of an actual lilac, he refines his L-system so that the figure it produces more closely resembles reality. Using his refined L-system, he generates the branching structure of the lilac. He then uses the same technique, with a different L-system, to produce the blossoms. The result? A lilac grows before his eyes. Not a real lilac, but a mathematical one, produced on a computer (see Figure 4.7).

What does one of Prusinkiewicz's computer-generated lilacs look like? It looks stunningly like a photograph of a real lilac. Just as the animal coat patterns produced by Murray's equations look just like the real thing.

Both pieces of work show how the complex shapes of nature can result from very simple rules. As Prusinkiewicz says: "A plant is repeating the

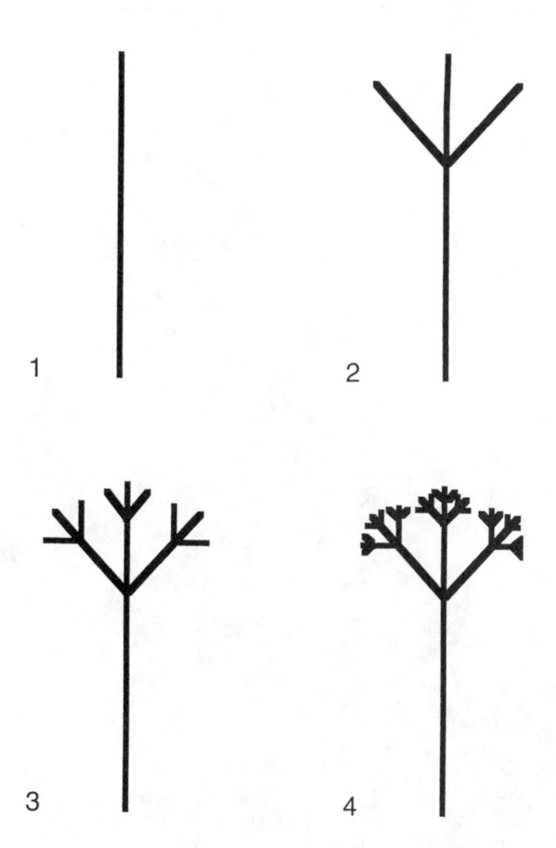

FIGURE 4.6 Repetition of a simple reproduction rule rapidly generates a tree-like structure. The rule is to add two new branches two-thirds of the way up each topmost branch, each one equal to one-third the length of that branch.

same thing over and over again. Since it is doing it in so many places, the plant winds up with a structure that looks complex to us. But it's not really complex; it's just intricate."

Murray and Prusinkiewicz both see a deeply satisfying aesthetic aspect to their work. For instance, Murray says:

> When I'm walking in the woods on my own or with my wife or daughter or son, and I have time to look around, I find it quite difficult not to

FIGURE 4.7 An electronic lilac generated on a computer by the iteration of some simple growth rules. Examples such as this demonstrate that much of the seeming complexity of the natural world can be produced by some very simple rules.

> look at a fern or the bark of a tree and wonder how it was formed—why is it like that? It doesn't mean one doesn't appreciate the beauty of a sunset or of a flower; the ideas or questions are interspersed with that sort of appreciation.

Prusinkiewicz has a similar sense of an enhanced appreciation of life:

> When you appreciate the beauty of plant form, it comes not only from the static structure, but often also from the process that led to the structure. To a scientist who appreciates the beauty of this flower or leaf, it is an important aspect of understanding to know how these things were evolving over time. I call it the algorithmic beauty of plants. It's a little bit of hidden beauty.

THE PATTERNS OF BEAUTY THE EYE CANNOT SEE

In geometry we study some of the visual patterns that we see in the world around us. Those visual patterns can be the "obviously mathematical" shapes studied by the ancient Greeks—triangles, circles, polyhedra, and the like—or the coat patterns of animals and the growth patterns of plants and flowers. (It is really a matter of definition whether you call these more recent studies "geometry." In any event, they are definitely mathematics, and they deal with visual patterns of shape.)

But our eyes perceive other patterns, patterns not so much of shape but of form. Symmetry is an obvious example. The symmetry of a flower or snowflake is clearly related to its geometric regularity. Yet we do not really see symmetry—at least, not with our eyes; rather, we perceive it, with our minds. The only way to "see" the actual patterns *of* symmetry (as opposed to symmetrical patterns) is through mathematics. By making visible the otherwise invisible patterns of symmetry that contribute to beauty, the study of symmetry captures one of the deeper, more abstract aspects of shape.

I shall go a bit deeper into this example than I did for animal coat patterns and flowers, and as a result most readers are likely to find the next few pages hard going. Please try to bear with me. As I develop my theme of investigating how human beings acquired the ability to do mathematics (and why many people seem unable to use that ability), I shall use this particular example to illustrate my argument.

The first step is to find a precise way of looking at symmetry—a way that allows the mathematician to start to use formal reasoning and (most likely) to write down formulas and equations. This initial step in developing a new branch of mathematics is often one of the most difficult. Not mathematically difficult. After all, there is no mathematics yet! Rather, the problem is finding a new way to look at the phenomenon.

What is symmetry? In everyday terms, we say an object (a vase, perhaps, or a face) is symmetrical if it looks the same from different sides or from different angles, or when it is reflected in a mirror. These general observations do not exhaust all the possibilities, but they capture the main idea.

First, we take that general idea of "looking the same from several (or many) angles" and put it into concrete form. (By starting with physical objects and viewing symmetry in terms of concrete manipulations of those objects, we will arrive at an abstract, formal definition of symmetry that can apply to totally abstract objects.)

Let's start by saying what we mean by "looking the same from a different angle." Imagine you have some object in front of you. It could be a two-dimensional figure or a three-dimensional object. Now imagine that the object is rotated about some line or point (see figures 4.8 and 4.9). Does the object *look* the same after the manipulation as it did before—are its position, shape, and orientation the same? If they are, we shall say that the object is "symmetrical" *for that particular manipulation*.

For example, if we take a circle and rotate it about its center through any angle we please, the resulting figure looks exactly the same as it started out (see Figure 4.8). We say that the circle is symmetrical for any rotation about its center. Of course, unless the rotation is through a full 360° (or a multiple of 360°), any point on the circle will end up in a different location. The circle will have *moved*. But, even though individual points have moved, the figure *looks* exactly the same afterward as it did before.

A circle is symmetrical not only for any rotation about its center but also for a reflection in any diameter. Reflection here means swapping each point of the figure with the one directly opposite with respect to the chosen diameter. For example, with a clock face, reflection in the vertical diameter swaps the point at 9 o'clock with the point at 3 o'clock, the point at 10 o'clock with the point at 2 o'clock, etc. (see Figure 4.9).

The circle is unusual, in that it has many symmetries—in everyday language, we would call it "highly symmetrical."

A square, on the other hand, has less symmetry than a circle. If we rotate a square through 90° or through 180° in either direction, it looks the same. But if we rotate it through 45°, it looks different—we see a diamond. Other manipulations of a square that leave it looking the same are

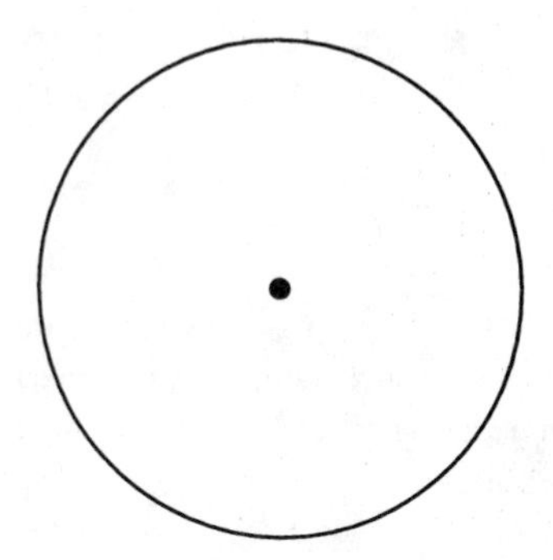

FIGURE 4.8 The circle looks exactly the same if it is rotated about the center through any angle or reflected in any diameter.

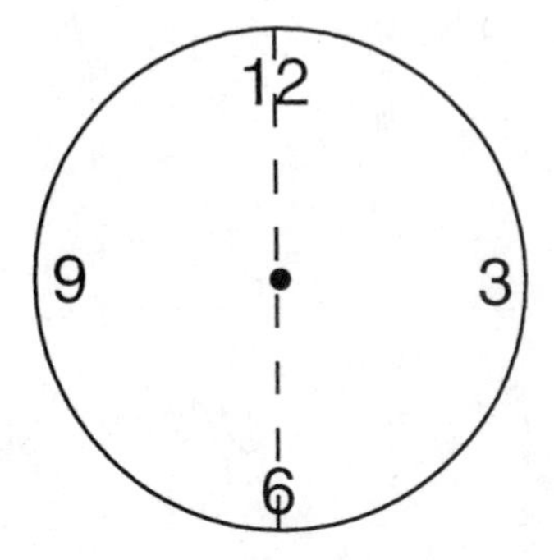

FIGURE 4.9 Reflecting a clock face in the diameter from 12 down to 6 swaps the 3 and the 9.

reflecting it about either of the two lines through the center point, parallel to an edge. Or we can reflect the square in either of the two diagonals. Figure 4.10 shows that each of these manipulations moves specific points of the square to other positions, but as with the circle, the figure we end with *looks* exactly the same—in position, shape, and orientation—as it did before.

A human face has symmetry, but not as much as a square. The face looks the same if it is reflected in a vertical line through the middle of the nose (i.e., swap left and right sides), a transformation that is easily achieved with a transparency photograph by simply flipping it left to right. But any other rotation or reflection produces a result that looks different. (I am assuming a perfect face here. In reality, there are no truly symmetrical human faces, just as no physical object is a perfect circle. Mathematics always studies imaginary perfect versions of real-world objects.)

In three dimensions, the human body is symmetrical for a reflection in the plane running vertically down the center of the body, front to back—the reflection that swaps the left side of the body with the right. This symmetry lies behind what at first seems a puzzling feature of mirrors. Namely, when you look in the mirror, you see yourself with your left and right sides swapped around (lift up your left hand, the reflection looking

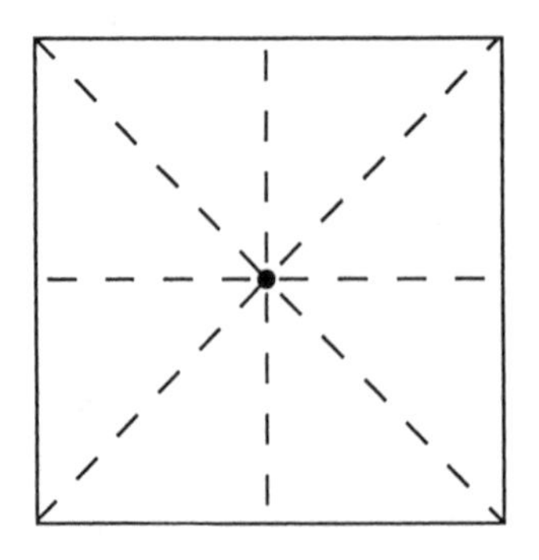

FIGURE 4.10 The square looks exactly the same after a rotation about the center through one or more right angles or a reflection in any of the dashed lines.

back at you lifts its right hand), and yet your reflection does not have top and bottom swapped. How is it that the mirror swaps left and right but not top and bottom? What happens when you lie down in front of the mirror, with your left side on top and your head on the right?

The explanation for this strange phenomenon is that a mirror does not in fact swap anything around. If you wear a watch on your left wrist, then the reflection has a watch on the wrist directly opposite your left one. Likewise, the head of the reflected figure is directly opposite your head. But because of the left–right symmetry of the human body, you see your reflection *as if it were another person facing you*. Another person facing you would indeed have his or her right hand opposite your left hand.

To continue, we have moved from an everyday idea of "symmetry" to the more precise notion of *symmetry with respect to a particular manipulation of the object*. The greater the number of manipulations that leave a figure or object looking the same (in position, shape, and orientation), the more "symmetrical" it is in the everyday sense.

Since we want to apply our concept of symmetry to things other than geometric figures or physical objects, we shall begin to use the word "transformation" rather than manipulation from now on. A *transformation* takes a given object (which may be an abstract object) and transforms it into something else. The transformation might simply be *translation* (moving the object to another position without rotating it), or it could be *rotation* (about a point for a two-dimensional figure; about a line for a three-dimensional object) or *reflection* (in a line for a two-dimensional figure; in a plane for a three-dimensional object). Or it could be something that is not generally possible for a physical object, such as stretching or shrinking.

The key to the mathematical study of symmetry is to look at transformations of objects rather than the objects themselves.

To a mathematician a *symmetry* of a figure is a transformation that leaves the figure invariant. "Invariant" means that, taken as a whole, the figure *looks* the same after the transformation, in terms of position, shape, and orientation, as before, even though individual points of the figure may have been moved.

Because translations are included among the possible transformations, the repeating wallpaper patterns we considered earlier are symmetries. In fact, the mathematics of symmetry is what lies behind the proof that there are exactly seventeen possible ways to repeat a particular local pattern. The permissible transformations—the "symmetries" for wallpaper patterns—have to work on the entire wall, not just one part of it. This restriction is what limits the number of symmetries to seventeen.

The proof of the wallpaper-pattern theorem involves a close examination of the ways transformations can be combined to give new transformations, such as performing a reflection followed by a counterclockwise rotation through 90°. It turns out that there is an "arithmetic" of combining transformations, just as there is an arithmetic (the familiar one) for combining numbers. In ordinary arithmetic, you can add two numbers to give a new number, and you can multiply two numbers to give a new number. In the "arithmetic of transformations," you combine two transformations to give a new transformation by performing one of the transformations followed by the other.

The arithmetic of transformations works in some ways like the arithmetic of numbers. But there are interesting differences. The discovery of this strange new arithmetic in the latter part of the eighteenth century opened the door to a host of dazzling new mathematical results that affected not only mathematics but physics, chemistry, crystallography, medicine, engineering, communications, and computer technology as well.

WHY EVEN MATHEMATICAL LONERS LIKE TO WORK IN GROUPS

When mathematicians talk about "groups," chances are they are not talking about encounter groups or hot tub parties, but about these powerful new kinds of arithmetic—which mathematicians call by the rather daunting name "group theory."

Our initial look at symmetry has taken us on a path of increased precision and deeper abstraction. Starting with an intuitive sense of things being "symmetrical," we were led to formulate a precise notion of a mathematical "symmetry"—a transformation that leaves the overall object or figure invariant, i.e., looking exactly the same. Then we realized that when we combine two such symmetries of a given object or figure, we get another symmetry (of that object or figure). That in turn reminded us of combining two numbers to give a new number in arithmetic, say by adding or multiplying them together.

The next step, which we'll carry out below, is to see how the *pattern* of combining symmetries compares with the *pattern* of adding numbers or the *pattern* of multiplying numbers.

Notice how, by remaining constantly on the lookout for new patterns (or even more, for familiar patterns in new guises), we keep changing our viewpoint. I am sure that this constant shifting of view and the accompanying steady increase in abstraction (from patterns, to patterns of patterns, to patterns of patterns of patterns) are among the features of advanced mathematics that most people find hardest to cope with. (I certainly do.)

For the eighteenth-century mathematicians who first trod it, the "symmetry path" we are following was very much a journey into the unknown. It took tremendous insight and breathtaking creativity to see that this simple idea of combining symmetry transformations would lead to a powerful new kind of arithmetic: the arithmetic of groups.

Given any figure, the *symmetry group* of that figure is the collection of all transformations that leave the figure invariant. For example, the symmetry group of the circle consists of all rotations about the center (through any angle, in either direction), all reflections in any diameter, and any combination of such. Invariance of the circle under rotations about the center is referred to as rotational symmetry; invariance with respect to reflection in a diameter is called reflectional symmetry. Both kinds of symmetry are recognizable by sight.

Since it is such a simple example, I'll use the symmetry group of a circle to show how you do "arithmetic" with a group. This will involve a bit of algebra. It's a bit like the algebra familiar to you from high school, except that whereas in high school algebra the letters x, y, and z generally denoted unknown numbers, here the letters S, T, and W denote symmetry transformations (of the circle in the first instance).

If you get totally lost, you can always skip ahead to the start of the next chapter. You will still be able to finish the book, although you will not be able to appreciate fully the case I make that mathematical thought is just a variation of other kinds of thought. (Of course, you can always come back later and take another look. By the book's end, I hope to have convinced you that you really do have the ability to follow such a discussion.) Here goes.

If S and T are two transformations in the circle's symmetry group, then the result of applying first S and then T is also a member of the symmetry group. (Why? Because, as both S and T leave the circle invariant, so too does the combined application of both transformations.) It is common to denote this double transformation by $T \circ S$. (This is read as "T composed with S." There is a good reason for the order here, where the transformation applied first is written second, having to do with an abstract pattern that connects groups and other branches of mathematics, but I shall not go into that connection here.)

As I mentioned, this method of combining two transformations to give a third is reminiscent of addition and multiplication, which combine any pair of whole numbers to give a third. To the mathematician, ever on the lookout for patterns and structure, it is natural to want to see which properties of addition and multiplication of whole numbers are echoed by the act of combining symmetry transformations.

First, the operation is what is called *associative:* if S, T, W are transformations in the symmetry group, then:

$$(S \circ T) \circ W = S \circ (T \circ W)$$

What this equation means is that if you want to combine three transformations, it doesn't matter which two you combine first (as long as you keep them in the same order: S, then T, then W). You can form $S \circ T$ first, and then combine the result with W, or you can form $T \circ W$ first and then combine S with that result. The answer you get will be the same in both cases. In this respect, this new operation is very much like addition and multiplication of whole numbers.

Second, there is an *identity* transformation that leaves unchanged any transformation it is combined with. It's the *null rotation,* the rotation through angle 0. The null rotation, call it I, can be applied along with any other transformation T, to yield:

$$T \circ I = I \circ T = T$$

The rotation I plays the same role here as the number 0 does in addition ($x + 0 = 0 + x = x$, for any whole number x) and the number 1 in multiplication ($x \times 1 = 1 \times x = x$).

Third, every transformation has an *inverse:* if T is any transformation, there is another transformation S such that the two combined together give the identity:

$$T \circ S = S \circ T = I$$

The inverse of a rotation is a rotation through the same angle in the opposite direction. The inverse of any reflection is that very same reflection. To obtain the inverse for any finite combination of rotations and reflections, you take the combination of backward rotations and re-reflections that exactly undoes its effect: start with the last one, undo it, then undo the previous one, then its predecessor, and so on.

The existence of inverses is another property familiar to us in addition of whole numbers: for every whole number m there is a whole number n such that:

$$m + n = n + m = 0 \text{ (the identity for addition)}$$

The inverse of m is just negative m. That is, $n = -m$.

The same is not true for multiplication of whole numbers. It is not the case that for every whole number m there is a whole number n such that:

$$m \times n = n \times m = 1 \text{ (the identity for multiplication)}$$

In fact, only for the whole numbers $m = 1$ and $m = -1$ is there another whole number n that satisfies the above equation. Whereas for the first two properties (associativity and the existence of an identity) group arithmetic for the circle is just like both addition and multiplication of whole numbers, for the third property (existence of inverses) group arithmetic is like addition of whole numbers but different from multiplication of whole numbers.

To summarize, any two symmetry transformations of a circle can be combined by the combination operation to give a third symmetry transformation, and this operation has the three properties familiar to many of us from ordinary arithmetic of whole numbers: associativity, identity, and inverses.

Although we were thinking about symmetries of a circle, everything we just observed is true for the group of symmetry transformations of any figure or object. (If you don't believe me, pick some other figure, say a square, and go back and check.) In other words, the arithmetic of symmetry transformations applies to any object or figure whatsoever. What is more, that arithmetic is like the addition (but not the multiplication) of whole numbers.

Once mathematicians realized that they had found a new kind of arithmetic, they immediately started to study it in a completely abstract fashion. Now, if you think things have become pretty abstract already, you're right. But abstraction is one of the most powerful weapons in mathematicians' armory, and it's one they use whenever the opportunity arises. In the case of groups, the next abstraction step is to forget the actual symmetries and the figure or object they are based on, and just look at the arithmetic. The result is a completely abstract definition of the (abstract) mathematical object called a "group."

In general, whenever mathematicians have some set, G, of entities (which could be the set of all symmetry transformations of some figure, but need not be) and an operation that combines any two elements x and y in the set G to give a further element $x * y$ in G (read as "x star y" or simply "xy"), they call this collection a *group* if the following three conditions are met:

G1. For all x, y, z in G, $(x * y) * z = x * (y * z)$.

G2. There is an element e in G such that $x * e = e * x = x$ for all x in G. (e is called an identity element.)

G3. For each element x in G there is an element y in G such that $x * y = y * x = e$, where e is as in condition G2.

These three conditions (generally called the *axioms*[2] for a group) are just the properties of associativity, identity, and inverses that we already observed for combining symmetry transformations of any figure. Thus the collection of all symmetry transformations of a figure is a group: G is the collection of all symmetry transformations of the figure, and is the operation of combining two symmetry transformations.

2 Perhaps the best-known example of axioms are the ones Euclid wrote down for plane geometry in his great work *Elements*. He began by stating two initial assumptions:

1. A point has no parts.
2. A line has no breadth.

and then presented five postulates which, he believed, were sufficient to prove all the true facts of plane geometry:

1. There is a line connecting any two points.
2. A line may be continued indefinitely.
3. Given any point, you can draw a circle of any radius with that point as its center.

It should also be clear that if G is the set of whole numbers and the operation * is addition, then the resulting structure is a group. As we observed, the same is not true for the whole numbers and multiplication. But you should have no trouble convincing yourself that, if G is the set of all rational numbers (i.e., whole numbers and fractions) apart from zero, and * is multiplication, then the result is a group. All you have to do is show that the conditions G1, G2, and G3 above are all valid for the rational numbers when the symbol * denotes multiplication. In that example, the identity element *e* in axiom G2 is the number 1.

The finite arithmetic we use to tell the time is another example of a group. In this arithmetic, the 12-hour clock has the whole numbers 1, . . . , 12 (which constitute the set G), and we add them according to the rule that we go back to 1 when we get to past 12. For example, 9 + 6 = 3, which looks a little odd until you remember that:

9 o'clock + 6 hours = 3 o'clock

For this arithmetic, the associativity condition (G1) is valid. For example:

(9 o'clock + 6 hours) + 2 hours = 9 o'clock + (6 hours + 2 hours)
= 5 o'clock

(Just work out each part for yourself). So:

$$(9 + 6) + 2 = 9 + (6 + 2) = 5$$

---

4. All right angles are the same.
5. Given any line and a point not on that line, there is a unique line through that point, parallel to the given line. (Two lines are parallel if they do not meet, no matter how far extended.)

In fact, Euclid missed a number of very subtle assumptions that he used unknowingly in his proofs, but by and large he got it right. The missing axioms were supplied by the great German mathematician David Hilbert in the late nineteenth century.

What about an identity? Well, in clock arithmetic, adding 12 takes us back to the same hour. For example,

$$2 \text{ o'clock} + 12 \text{ hours} = 12 \text{ o'clock} + 2 \text{ hours} = 2 \text{ o'clock}$$

So

$$2 + 12 = 12 + 2 = 2$$

So G2 holds, with the number 12 being the identity. (Note that clock arithmetic does not have 0.) What about condition G3? To get the inverse for any number in G, you simply continue around the clock face until you reach 12. For example, the inverse of 7 is 5, because

$$7 + 5 = 5 + 7 = 12$$

and 12 is the identity in 12-hour clock arithmetic. (Thus, in 12-hour clock arithmetic, we get the rather strange-looking equation $-7 = 5$.)

Clock groups are cute, but not particularly interesting to a mathematician. My purpose in mentioning them was simply to show you that the group concept can arise in many different contexts besides the ones we've looked at previously. In fact, groups arise all over the place. The group "pattern" can be found hidden in all sorts of phenomena. This is why the mathematics of groups is taught to every university student of mathematics.

NOW FOR THE REALLY HEAVY STUFF

Before looking at one further example of a symmetry group, it is worth spending a few moments looking at the three conditions that determine whether a given collection of entities and an operation constitute a group.

The question is: what other properties of groups follow automatically from the three group axioms? Anything that we can show to be a logical consequence of the axioms will be automatically true for any particular group.

(This is probably another point where some readers will want to bail out and skip to the next chapter. As I said before, the longer you stick with this chapter, the better you will be able to appreciate my arguments about the nature of mathematical thought. But again, you can always come back later.)

The first condition, G1, the associativity condition, is already very familiar to us in the case of the arithmetic operations of addition and multiplication (although not subtraction or division). I won't say anything more about that.

Condition G2 asserts the existence of an identity element. In the case of addition of whole numbers, there is only one identity: the number 0. Is this true of all groups, or is it something special about whole number arithmetic?

In fact, any group has exactly one identity element. If e and i are both identity elements, applying the G2 property twice in succession gives the equation:

$$e = e * i = i$$

So e and i must be one and the same.

This last observation means in particular that there is only one element e that can figure in condition G3. (G3, remember, says that for each element x in G there is an element y in G such that $x * y = y * x = e$.) Using that fact, we can go on to show that, for any given element x in G, there is only *one* element y in G that satisfies the condition in G3. This is a bit harder to prove. Hang on.

Suppose y and z are both related to x as in G3. That is, suppose that:

$$x * y = y * x = e$$
$$x * z = z * x = e$$

Then:

$$\begin{aligned} y &= y * e \text{ (by the property of } e) \\ &= y * (x * z) \text{ (by equation 2)} \\ &= (y * x) * z \text{ (by G1)} \\ &= e * z \text{ (by equation 1)} \\ &= z \text{ (by the property of } e) \end{aligned}$$

So y and z are one and the same. In other words, there is only one such y for a given x.

Since there is precisely one y in G related to a given x as in G3, that y may be given a name: it is called the (group) *inverse* of x and is often denoted x^{-1}. And with that, I have just proved a theorem in group theory: in any group, every element has a unique inverse. I proved this by deducing it logically from the group axioms, the three initial conditions, G1, G2, G3.

For most people, the above algebra is already a stretch. But for a mathematician, it is fairly straightforward. (The reason for this difference in comprehension is, of course, one of the things this book sets out to explain.) Whether you find it hard or easy, however, it does illustrate the enormous power of abstraction in mathematics. There are many, many examples of groups in mathematics. Having proved, *using only the group axioms*, that group inverses are unique, we know that this fact applies to every single example of a group. No further work is required. If tomorrow you come across a quite new kind of mathematical structure, and you determine that what you have is a group, you will know at once that every element of your group has a single inverse. In fact, you will know that your newly discovered structure possesses *every* property that can be established—in abstract form—on the basis of the group axioms alone.

The more examples there are of a given abstract structure, such as a group, the wider the applications of any theorems you prove about it. The cost of this greatly increased efficiency is that one has to learn to work with highly abstract structures—with abstract patterns of abstract entities.

In group theory, it seldom matters what the elements of a group are or what the group operation is. Their nature plays no role. The elements could be numbers, transformations, or other kinds of entities, and the operation could be addition, multiplication, composition of symmetry transformations, or whatever. All that matters is that the objects together with the operation satisfy the group axioms G1, G2, and G3.

One final remark concerning the group axioms is in order. Anyone familiar with the commutative laws of arithmetic might well ask why we did not include it as a fourth axiom:

G4. For all x, y in G, $x * y = y * x$.

The absence of this law meant that in both G2 and G3, the combinations had to be written two ways. For instance, both $x * e$ and $e * x$ appear in G2. If commutative law were true, we could have written G2 as:

There is an element e in G such that $x * e = x$ for all x in G.

The reason mathematicians do not include an axiom G4 is that it would exclude many of the examples of groups that mathematicians wish to consider. By writing G2 and G3 the way they do, and leaving the commutative law out, the group concept has much wider application than it otherwise would.

Consider, for example, a symmetry group a bit more complicated than a circle—the equilateral triangle shown in Figure 4.11. This triangle has precisely six symmetries. There is the identity transformation, I (the transformation that makes no changes whatsoever), clockwise rotations v and w through 120° and 240°, and reflections x, y, z in the lines X, Y, Z, respectively. (The lines X, Y, Z stay fixed as the triangle moves.) There is no need to list any counterclockwise rotations, since a counterclockwise rotation of 120° is equivalent to a clockwise rotation of 240° and a counterclockwise rotation of 240° has the same effect as a clockwise rotation of 120°.

There is also no need to include any combinations of these six trans-

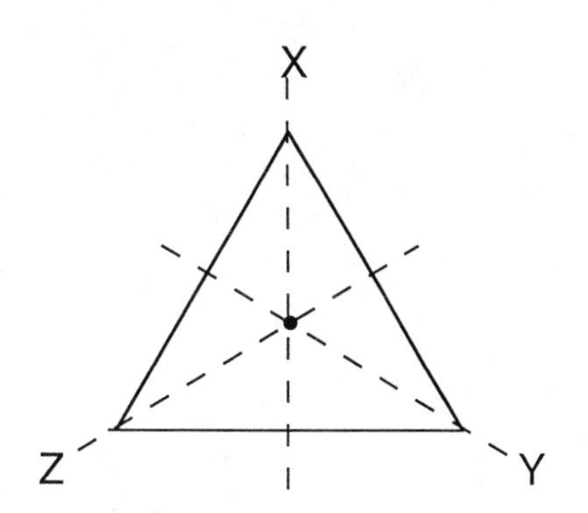

FIGURE 4.11 The triangle looks the same if it is rotated about the center through 120° in either direction or if it is reflected in any of the dashed lines marked X, Y, Z.

formations, since the result of any such combination is equivalent to one of the six given. The table shown in Figure 4.12 gives the basic transformation that results from applying any two basic transformations. For instance, the transformations x and v combine to give y, which we write as:

$$x \circ v = y$$

Again, the result of applying first w and then x, namely the group element $x \circ w$, is z, and the result of applying v twice in succession, $v \circ v$, is w. The group table also shows that v and w are mutual inverses and x, y, z are each self-inverse. Moreover, $x \circ y = v$ and $y \circ x = w$, so this group is not commutative.

Since the combination of any two of the given six transformations is another such transformation, it follows that the same is true for any finite combination. You simply apply the pairing rule successively. For example, the combination $(w \circ x) \circ y$ is equivalent to $y \circ y$, which in turn is equivalent to I.

And there you have the group theorist's equivalent of the multiplication table that caused you so much trouble in elementary school. For all the abstraction, we have been doing nothing more complicated than elementary school arithmetic.

So why does it feel much more complicated?

o	I	v	w	x	y	z
I	I	v	w	x	y	z
v	v	w	I	z	x	y
w	w	I	v	y	z	x
x	x	y	z	I	v	w
y	y	z	x	w	I	v
z	z	x	y	v	w	I

FIGURE 4.12 Multiplication table for the symmetry group of an equilateral triangle.

Partly because it seems so much more abstract. Yet I don't think combining transformations is any more abstract than combining numbers. The only difference is that transformations are *processes* you *perform* whereas numbers are based on *collections* that you *count*. And that's not much of a difference. Both can be done concretely—with cardboard cutout figures for symmetry transformations and with collections of counters for arithmetic. Maybe if we were taught the "arithmetic" of symmetry transformations as young children, and encountered numbers only when we were older, we would find groups easier than arithmetic.

This points to a significant factor affecting the way we learn mathematics. Young children have not only a great ability to master abstraction, but an instinct to do so. For that is precisely what is involved in learning to use language, which all young children do with ease. Since children are taught arithmetic at a very young age, they achieve a reasonable mastery of it. By the time they are faced with learning a "different arithmetic"—be it algebra at high school or group theory in college—not only have they lost the ability to master abstraction spontaneously, but even worse, most have also developed an expectation that they *cannot* master it. And, as with most things in life, we tend to find what we expect.

·5·

DO MATHEMATICIANS HAVE DIFFERENT BRAINS?

THE STRANGE CASE OF EMILY X

AMERICAN READERS in their fifties and older will probably be able to identify the person I am calling "Emily X." In the 1960s, Emily was a brilliant and vivacious mathematics student at the University of California at Berkeley. Both her parents were physicians at the nearby Oakland Children's Hospital, and she had a younger brother, Todd, then still in high school. Her mathematical ability was so great that as a sophomore she was already taking graduate classes in addition to her undergraduate courses, and her professors had begun steering her toward graduate school. Her future seemed assured.

Then, in her junior year, Emily disappeared. There was no note, no message to her friends. Her parents and younger brother were devastated.

Days turned into weeks and months, and eventually the newspapers dropped the story. A year went by, then two, then three, and still there was no clue as to her whereabouts. Most people who knew her feared the worst—that she had been abducted and murdered.

Then, almost five years to the day after she vanished, Emily reappeared. She simply walked into her former home, poured a glass of milk, made a peanut butter and jelly sandwich, switched on the television, and sat down to wait for her parents to come home from work.

She looked fit and healthy, she showed no signs of any physical abuse, and she was outwardly happy. But she had absolutely no recollection of having been away, or indeed of the preceding five years. She did not know that President Kennedy had been shot. She had never heard of the Beatles. She thought it was still 1962.

The only apparent change was one that only her former mathematics professors could spot. Emily's mathematical ability had increased dramatically. Five years earlier, she had been a very promising sophomore. Now she was on a par with the members of Berkeley's world-famous mathematics faculty.

How could this have happened? It soon became apparent that she had not attended any other university. Even if she had studied in a foreign country, as some commentators suggested, it could not account for her total inability to remember anything but the mathematics she had learned.

It took Emily about six months of intense effort to fill in the missing years. During this time, she resumed her studies. Or rather, she worked alongside her former teachers as they tackled major unsolved problems of mathematics.

Then, after she had "caught up" with the world, two major events occurred that again changed her life.

First, she woke up one morning and found she had apparently lost all her mathematical ability. She could still do arithmetic, and she could carry out the symbolic manipulations of elementary algebra as well as any bright high school student. But she could not follow even the simplest mathematical proof—it was as if she did not understand what a proof was.

Second, that night, she experienced the first of many mental flashbacks that would eventually allow her to piece together her experiences during what she called her "lost years."

Thus far, everything is well documented. The doubts occur when Emily tells of her life during her five "lost years." Her descriptions of life in "a cold place, with snow everywhere, and a sky of shining silver" (from her autobiography, *My Lost Years*) are so detailed, and so logically consistent, that it is hard to imagine that they are fabricated. Yet she describes a world most of us would dismiss as science fiction.

Although Emily eventually recovered sufficiently to live an outwardly normal life, she never married or formed any close personal relationships. This is particularly significant, given that her descriptions of her lost years are filled with highly detailed accounts of friendships, love affairs, weddings, and marriages. Only much later, when she had retrieved the more painful memories from her lost years, did it become clear what had led to this interest.

Some of the interpersonal relationships in Emily's other world are straightforward enough, others seem strange. For instance, in one section she describes (in extreme detail) her "other world" friends Janet and Eric, a married couple, and how they fell in love with, and then married, a young man called Paul. Three-way marriages were apparently quite normal in Emily's other world. In her own words (*My Lost Years*, p. 193):

> The law does not prevent marriages of any size, but financial concerns do. In a three-way marriage, for instance, the new partner has exactly the same rights as the first two partners. When Paul married Janet and Eric, he had the same rights as they did. It would have been just the same if Eric had married Paul first and then they had married Janet. Being the first in a marriage offers no legal advantages. Couples definitely give something up if they marry a third person. Of course, they gain a lot as well. But most decide not to.

Elsewhere, Emily describes her own "empty marriage." The term turns out to have a different meaning in Emily's other world than it has in the all-too-familiar world of failed marriages. An "empty marriage," Emily explains, was a marriage ceremony in which a person married a state-appointed individual called a "tolen." Although legally valid, the ceremony does not affect either party's legal status. Its only purpose is to enable the individual to experience a marriage ceremony.

Many American readers will remember Emily's appearance on *The Tonight Show* with Johnny Carson. Carson recounts the interview in his own autobiography (*Here's Johnny*, p. 207):

> "What about divorce?" I asked at one point in the conversation.
>
> "Oh yes, of course there's divorce," Emily replied.
>
> "Maybe you were living in Nevada!" I quipped. She did not respond to my humor, so I asked another question: "Tell me, how do you go about getting a divorce in your world? Is it easy?"
>
> "No, it's not easy at all," Emily replied. "Divorce is just a special kind of marriage. You have to find your spouse's nullifier. Then you go through a normal marriage ceremony with the nullifier. Then you are no longer married."
>
> I remarked that this didn't sound difficult at all, and joked that Elizabeth Taylor did this all the time, but Emily ignored my attempt at humor. She went on to explain the process step by step, as if she were telling a child how to boil an egg. "Everyone has a nullifier," she said, "but you only have one of them, so the difficult thing is to find him. It can take a lot of time. Maybe you'll never find him. That can happen. Of course, there are agencies that specialize in locating nullifiers . . . "
>
> By now the producer was waving furiously at me to liven things up, and I had to prevent the interview from turning into a lecture. "I'll bet they are expensive," I broke in again.
>
> Unfortunately, once again Emily did not pick up on my attempt to inject a bit of humor. She simply confirmed, in a very matter-of-fact way,

that divorces were both expensive and difficult to arrange. The producer was getting desperate. I tried again:

"Then I guess you were not living in Nevada after all!"

This time it worked. She picked up on my intention. "No, I'm sure it wasn't Nevada," she replied with a laugh. It was a great laugh.

At that point, the show came to an end. Emily's final laugh left a positive impression on the audience. She could have come across sounding like a freak, and she almost did, but in the end she didn't. That last remark saved her. And the show.

Looking back, it was one of my favorite shows. Emily was a fascinating person. I don't have any record of the rest of our conversation, but it was probably as long as I have spent with a guest after we had gone off the air.

But if Emily had not spent her lost years in Nevada, where had she been? Many dismiss Emily's story as the delusions of a schizophrenic. But they cannot explain the fact that for five years not a single person saw her or heard from her, and that during that same period, she went—untutored by any human being—from being a bright mathematics undergraduate to a world-class research mathematician. You can't fake mathematics.

In the next section, I'll give my own explanation of the Emily X case. It may well change the way you think about mathematics.

HOW DO MATHEMATICIANS THINK?

Is it significant that Emily was a mathematician? Do mathematicians think differently from other people? Do they have different brains? The answers to these questions are, respectively, yes, yes *in a way*, and no. Before I explain these answers, here's a simple test of your logical reasoning skills. (Actually, it's only simple to state. Most people find it extremely hard, and most people get the answer wrong.)

Imagine I lay four cards on a table in front of you. I tell you that each card has a number on one side and a letter on the other. On the uppermost faces of the cards you see the four symbols:

E K 4 7

I then tell you that the cards are printed according to the rule: *If a card has a vowel on one side, it has an even number on the other side.*

Your task: which cards do you *have* to turn over to be sure that all four cards satisfy this rule?

Okay, before you read any further, try to answer the above question yourself.

Have you done that? Then let's see how you did.

I am pretty sure you decided to turn over the card with the vowel—the E. You undoubtedly realized that, if the number on the other side turned out not to be even, you would have disproved the rule. Almost everybody gets this card right. Likewise, most people realize that they needn't turn over the card with the consonant, K, since the rule says nothing about cards with consonants. So far so good. That leaves the two cards that show numbers. You might have decided to turn over the card with the 4. Some people think you must turn over this card, others don't. In fact you don't need to. If you turn it over and there is a vowel on the other side, that's in accordance with the rule; on the other hand, if you find a consonant on the other side, that too is consistent, since the rule says nothing about cards with consonants. Finally, you almost certainly did not see any point in turning over the card with the odd number, the 7. And yet it is crucial that you do: if there is a vowel on the other side, the rule would be broken. You have to check to eliminate that possibility.

In other words, you should turn over the E and the 7. This becomes a little easier to see if you look at the task a slightly different way. The only combination that goes against the rule is a vowel coupled with an odd number. So to see if the four cards satisfy the rule, you have to check that this forbidden combination does not occur. That means looking at cards bearing a vowel or an odd number—in the case at hand, an E or a 7.

This test is well known in the cognitive psychology community as the Wason test, after the British psychologist Peter Wason, who introduced it in the early 1970s. Most people get it wrong.

Now, to restore your confidence in your reasoning ability, here's a much simpler problem. You are in charge of a party where there are young people. Some are drinking alcohol, others soft drinks. Some are old enough to drink alcohol legally, others are under age. You, as the organizer, are responsible for ensuring that the drinking laws are not broken, so you have asked each person to put his or her photo ID on the table. At one table are four young people who may or may not be over the legal drinking age. One person has a beer, another has a Coke, but their IDs happen to be face down so you can't see their ages. You can, however, see the IDs of the other two people. One is under the drinking age, the other is above it. Unfortunately, you are not sure if they are drinking 7-Up or vodka and tonic. Which IDs and/or drinks do you need to check to make sure that no one is breaking the law?

Almost everyone can answer this question correctly. You check the ID for the person drinking the beer and you sniff the drink of the person whose face-up ID shows she is under age.

"So what's the big deal?" you ask. "This is a really simple problem, much easier than the one with the four cards." What's my point?

My point is that this is exactly the same problem as the first one! Here is the correspondence between the two problems:

Has a vowel	Is drinking alcohol
Has a consonant	Has a soft drink
Has an even number	Is above the legal drinking age
Has an odd number	Is below the legal drinking age

The law that *If you are drinking, you must be above the legal age* corresponds to our earlier rule, *If a card has a vowel on one side it has an even number on the other side.* Checking the beer drinker and the identifiably under-age person at the party corresponds to turning over the E and the 7 in the card problem.

So why do most people find the first problem so hard and the second one so easy? From a logical point of view, they are one and the same problem. The difference must be the presentation: what the problems are portrayed as being about. In one case the problem is an artificially contrived task that bears no relation to the real world. In the second case, the problem involves familiar, concrete objects and circumstances: young people, parties, and drinking laws. This is what makes all the difference.[1]

The Wason test is famous because the outcome is so dramatic. A change in presentation turns a difficult problem that most people get wrong into an extremely simple one that most easily get right. The fact is that people reason much better about familiar, everyday objects and circumstances than they do about abstract objects in unfamiliar settings, even if the logical structure of the task is the same.

With that in hand, let's return to Emily X.

What were your responses to Emily's story? Almost certainly, your interest was aroused initially on a human level: you wondered what had happened to Emily during those five "lost years."

If you are among the vast majority of readers, that will be the extent of your interest (apart perhaps from trying to remember the actual events if you lived in California during the early 1960s). If you are an anthropologist, a sociologist, or a psychologist, however, you will also have been interested in the marriage practices and laws of Emily's "other world." (If you do not have such a specialist's interest, then most probably you skimmed over her descriptions, finding them somewhat boring, just as

1 I should point out that it's the abstractness of the *problem structure* that makes the major difference in how well people perform on the Wason test. Posing the problem with the aid of real cards makes it no easier than if the problem is simply *described*, as I have done here. Nor is the problem any easier if it is stated in terms of familiar objects and activities, but with an unfamiliar logical structure, such as: "If a person eats hot chili, then he drinks cold beer."

Johnny Carson and his producer thought they were for *The Tonight Show* audience.)

For some readers then, the Emily story offers a second layer of interest. It's not that these readers do not respond to the story in a normal human fashion. Rather, their professional training and interest enable them to appreciate a second layer that is glossed over by most readers.

If you are a mathematician, you most probably had quite a different secondary reaction to Emily's descriptions. Almost certainly, you smelled a rat.

Okay, it's owning-up time: the Emily story is a fabrication. I simply made it up. But I did so in order to achieve a number of aims, so I hope you will forgive the deception. One aim of the story was to drive home the lesson of the Wason test: that you can easily understand mathematical concepts, provided they are presented in a familiar way.

What will have tipped off mathematical readers to the deception was Emily's descriptions of marital practices. Those three laws—the equity in three-party marriages, the provision of "empty marriage" ceremonies that do not result in an actual marriage, and the strange way of obtaining a divorce (finding and marrying a "nullifier")—are precisely the three axioms for a group (G1, G2, and G3) given in Chapter 4 (see page 103). But instead of presenting them in algebraic language, as I did then, Emily stated them in English. And instead of talking about "symmetry transformations" or "abstract group elements," Emily gave them as rules for marriage. But at heart they are exactly the same. In the terminology of the previous chapter, the marriage laws in Emily's "other world" mean that her other world is an example of a mathematical group.

One consequence of this last observation is that we can apply to Emily's world the theorem about abstract groups at the very end of Chapter 4, to *prove* that, as Emily claimed on *The Tonight Show*, each person in Emily's world has exactly one marriage nullifier.[2]

2 Technical aside for mathematical purists: Strictly speaking, the objects that form a group in the Emily example are not the individual people but the married families, and to make everything

Now we are starting to close in on what is involved in mathematical thinking.

THE FOUR LEVELS OF ABSTRACTION

One characteristic feature of the human brain that no other species seems to possess is the ability to think about abstract entities. Many species seem able to reason, if only in a very rudimentary way, about real objects in their immediate environment. Some, including chimpanzees and apes, seem to have an additional ability. A bonobo ape, for example, can carry out very limited reasoning about a single, real object it is familiar with but which is not currently present. The range of human thought, in contrast, is so broad as to constitute a different kind of activity altogether. We can think about practically anything we want: real objects we are familiar with but which are not in our immediate environment; real objects we have never seen but have simply heard or read about; or purely fictitious objects. Thus, whereas a bonobo ape may reason about how to retrieve a banana it just saw its trainer hide, we may think about a six-foot-long, gold-plated banana pulled along by two pink unicorns.

How is it possible to think about something that does not exist? To put it another way, just *what* is the object of our thought when we are thinking about, say, a pink unicorn? This is one of those questions about which philosophers argue endlessly, but the standard answer is that the objects of our thought processes are *symbols* (i.e., things that stand for or denote other things). We'll look at this idea in some detail in Chapter 8, but for now let me observe this: the symbols that form the object of the thoughts

---

work properly as a group, each unmarried person has to be regarded as constituting a "married family" consisting of just a single individual married to him- or herself. The reason for this is that, in a group, the product or combination of any two objects must be another object of the same kind. This kind of maneuver is often necessary when a mathematician seeks to apply a mathematical concept—in this case the group concept—to some real-world situation (even an imaginary real-world situation such as Emily's other world).

of an ape or a chimpanzee are restricted to symbolic representations of real objects in the world. On the other hand, the symbols that form the objects of our thoughts may also represent imaginary versions of real objects, such as imaginary bananas or imaginary horses, or even wholly imaginary objects put together from symbolic representations of real objects in the world, such as a gold banana or a unicorn.

I find it helpful to view abstract thought in terms of four levels.

Level 1 abstraction is where there is really no abstraction at all. The objects thought about are all real objects that are perceptually accessible in the immediate environment. (However, thinking about objects in the immediate environment might well involve imagining them moved to different locations in the environment, or arranged in different ways in the environment. Thus, I think it is reasonable to view this process as one of abstract thought, even though the objects of that thought are all concrete objects in the immediate environment.) Many species of animal seem capable of level 1 abstraction.

Level 2 abstraction involves real objects the thinker is familiar with but which are not perceptually accessible in the immediate environment. Chimpanzees and apes seem capable of thought at level 2 abstraction.

As far as we know, only humans are capable of *level 3 abstraction*. Here, the objects of thought may be real objects that the individual has somehow learned of but has never actually encountered, or imaginary versions of real objects, or imaginary variants of real objects, or imaginary combinations of real objects. Although objects in level 3 abstraction are imaginary, they can be described in terms of real objects—for example, we may describe a unicorn as a horse with a single horn on its forehead. As I shall explain in Chapter 8, the ability to think at level 3 abstraction is, to all intents and purposes, equivalent to having language.

Level 4 abstraction is where mathematical thought takes place. Mathematical objects are entirely abstract; they have no simple or direct link to the real world, other than being abstracted from the world in the sense exemplified by the groups discussed at the end of Chapter 4.

I can now propose one of the central tenets of my account of how the human brain evolved to the point where it could engage in mathematical thought. (I'll provide a supporting argument in Chapter 8.) It did not acquire new thought processes so much as apply existing thought processes to objects at a higher level of abstraction. In other words, the crucial development was one of increasing abstraction, not a greater complexity of thought processes.

As I have indicated, I'll provide a substantial argument to back up this claim in due course, including an examination of the way the brain works as a physical (electrochemical) device. In the meantime, let's see how this picture relates to some actual examples of mathematical thinking.

Remember the Wason test? The puzzle about a group of imaginary young people at a fictitious party involves level 3 abstraction. In contrast, the puzzle about the four cards involves level 4 thought. For, while cards with letters and numbers on them are at most level 3 abstractions, the letters and numbers are symbolic entities, and the problem scenario is accordingly level 4. This is true even if the problem is presented to you with real, physical cards. The fact that the puzzle involves cards with letters on them is not relevant to the way you try to solve the problem; it's essentially a problem in pure logic. Your thought processes would be the same if the letters and numbers were displayed on a computer screen and you were asked which keys to depress. (In other words, the card puzzle is to all intents and purposes entirely mathematical.) Your solution to the party problem, on the other hand, depends heavily on its being about young people and drinking laws. It is precisely in those terms that you solved the problem. This is why, even when people have been shown its relation to the four cards puzzle, they still have difficulty solving the card problem. The ease of the party problem does not transfer to the logically equivalent card problem.

In Chapter 8, I shall examine the way the human brain acquired the ability for level 3 abstraction. I will suggest that, in essence, the brain acquired the ability to simulate (in a symbolic fashion) the real world to a sufficient extent that the circuits that evolved to enable it to negotiate and survive in the world became able to operate without direct input from the world. (In physical terms, the brain acquired the ability to create activation patterns that resemble those created by direct sensory input from the world.)

Thus the mental process we go through in solving the party problem while sitting at home reading a book is little different from what we would go through if we really were faced with four young people at a party.

The key to being able to think mathematically is to push this ability to "fake reality" one step further, into a realm that is purely symbolic—level 4 abstraction. Mathematicians learn how to live in and reason about a purely symbolic world. (By "symbolic world" I don't mean the algebraic symbols that mathematicians use to write down mathematical ideas and results. Rather, I mean that the objects and circumstances that are the focus of mathematical thought are purely symbolic objects created in the mind.) Although it surely does not require a different kind of brain to deal with this world, it does involve considerable mental effort. All mathematicians *can* solve the four cards problem if they put their minds to it. But like everyone else, they find it harder than the party problem.

LIVING AMONG SYMBOLS

What is the nature of mathematical thought? What does it feel like to inhabit this symbolic world? Although I have been a mathematician for over thirty years, I am still not clear exactly what the nature of the mathematical thought process is.

I am sure it is not linguistic. Mathematicians do not think in sentences, at least not most of the time. The precise logical prose you find in mathematical books and papers is an attempt to *communicate* the results of mathematical thought. It rarely resembles the thought process itself.

I am in remarkably good company in having this view of mathematical thinking. For instance, in 1945, the distinguished French mathematician

Jacques Hadamard published a book titled *The Psychology of Invention in the Mathematical Field,* in which he cited the views of many mathematicians on what it feels like to do mathematics. Many of them insist that they do not use language to think about mathematics. Albert Einstein, for instance, wrote:

> Words and language, whether written or spoken, do not seem to play any part in my thought processes. The psychological entities that serve as building blocks for my thought are certain signs or images, more or less clear, that I can reproduce and recombine at will.

Hadamard himself makes the same point:

> I insist that words are totally absent from my mind when I really think ... even after reading or hearing a question, every word disappears the very moment that I am beginning to think it over.

Of particular interest are the mathematicians' descriptions of the way they arrived at the solutions to problems they had been working on. Time and again, the solution came at a quite unexpected moment, when the person was engaged in some other activity and was not consciously thinking about the problem. Moreover, in that inspirational moment the whole solution suddenly fell into place, as if the pieces of a huge jigsaw puzzle had been dropped onto the floor and miraculously landed as a complete picture. The mathematician "saw" the solution and instinctively knew it was correct.

No language is involved in this process. Indeed, with a problem for which the solution is fairly complex, it might take the mathematician weeks or even months to spell out (in linguistic form) the step-by-step logical argument that constitutes the official solution to the problem—the *proof* of the result.

So if mathematical thought is not linguistic, if mathematicians do not think in words (or algebraic symbols), how exactly does it *feel* to a mathematician thinking about mathematics? I can tell you how it feels to me, and from talking to other mathematicians I believe it feels the same to them—but given the subjective nature of the question, I cannot be certain.

When I am faced with a new piece of mathematics to understand or a new problem to work on, my first task is to bring to life the mathematical concepts involved. It is as if I have been given detailed instructions (including plans and blueprints) to build and furnish a house. By studying the instructions, I can locate and acquire the necessary materials, fittings, and furnishings, and, step by step, I construct the house. When it is finished, I move in. Because I have built the house myself, I know the layout. Although it feels a bit strange at first, within a few days I know the house so well that I can get around quite easily in the dark.

Of course, some things in the house were bought as completed units—the boiler, for instance—and I do not fully understand how those items are constructed or how they work. For example, I simply know the manufacturer's specifications of the boiler and how it fits into the heating system. I have neither the time nor expertise to learn boiler design and manufacture. I rely on the claims made by those who have designed and manufactured the boiler that it works as they say it does. If there later turns out to be a problem, I will have to call them in for help.

But I can live in the house without having to think consciously about the building process. The house is there, I am living in it, and I know my way around. My next task is to move the furniture around so that it best suits my lifestyle. Because I am familiar with the house and its contents, I do not have to think about each piece of furniture individually. I can concentrate on the *arrangement* of the furniture, what items go with what other items.

Similarly, when I start to think about a new piece of mathematics or am faced with a new mathematical problem, my first task is to build the "house"—a "house" built of abstract mathematical objects, fastened together by abstract logical and structural relationships. *Understanding* mathematics is like building the house and thereafter knowing my way around it.

These days, there is such a great volume of mathematics, much of it so complex, that no one mathematician could possibly master more than a small fraction. (More mathematics was produced in the twentieth century than in the entire 3,000-year history of mathematics prior to then.) Consequently, like every other mathematician, I have to take a lot on trust. For example, to solve a particular problem I may use results obtained by other mathematicians. In order to develop my solution, I must understand those results and be able to apply them to my own problem. But I may well have neither the time nor the necessary expertise to understand just how those other mathematicians obtained those results. I rely on their expertise, together with the fact that, in order to be published, their work had to be examined in detail and certified as correct by other mathematicians—just as when I purchase a sophisticated, computer-controlled boiler for my house, all I need to know are the power ratings and the installation procedures. I do not need to understand how the boiler works. I rely on the expertise of the designers and manufacturers who produced the device, together with the government certification procedures that ensure its safety.

Thus, understanding the mathematics required to solve my problem may well involve learning a number of mathematical facts without knowing exactly how they were established. Just how much I need to know about those supporting facts in order to solve my problem is a matter of mathematical judgment. I may find out later on that I need to learn more about how a particular fact was established by someone else—perhaps because I want to modify that earlier result in some fashion. But I won't worry about that possibility until it arises.

Once I have understood the mathematics involved in a particular problem, then I can try to solve it. *Trying to solve a problem* is like moving the furniture around in the house to find the best arrangement.

At least, that's how it felt when I was starting out in mathematics. Having been a mathematician for thirty years, I have now built a fair number of mathematical houses and know my way around them pretty well. These days, for the most part, I live in those old, familiar houses that I built long ago. I no longer build new houses. Tackling a new problem is now more like rearranging the existing furniture to accommodate a new acquisition. (To make sure that I do not have to spend half my life building new houses, I am careful to pick problems that can be solved using mathematical techniques I am familiar with. In other words, like all mathematicians these days, I *specialize* in just a few areas.)

Notice that, once the house has been built and the instructions and plans have been stored away, there is no more need for language. I simply *live* in the house. Language is required only if some problem sends me back to the plans, or if I want to remodel or purchase a new item. And, of course, I need language if I want to describe to someone else how I built the house or why I arranged the furniture the way I did.

To me, then, *learning* new mathematics is like constructing a mental house in my mind; *understanding* that new mathematics is like becoming familiar with the interior of my mental house; and *working on a mathematical problem* is like arranging the furniture. *Thinking* mathematics is like *living* in the house. As a mathematician, I create a symbolic world in my mind and then enter that world.

And that brings me back to Emily X. I made Emily a mathematician because doing mathematics involves entering another world—a world within the mind. Emily's "other world" is her symbolic world of mathematics. One of the things that enables her to be such a good mathematician is that, when she enters that other world, she becomes unaware of what is going on around her in the real world. I'll have more to say on this later in this chapter.

SOME CAN, SOME CAN'T—OR CAN THEY?

Why then do many people find mathematics completely incomprehensible while others handle it with seeming ease?

When we enter a house for the first time, we of course find it unfamiliar. By walking around for a while, however, looking into various rooms and peering into cupboards, we quickly get to know it. But what if we cannot enter the house, and our only knowledge of it comes from the instructions and plans that were used to build it? Moreover, what if those instructions and plans are written in a highly technical language that we find intimidating and incomprehensible? What if, try as we may, we cannot form any mental picture of the house? Then we are not going to get much of a sense of what it is like to live there. We are not going to be able to enter the house even in our imagination.

Notice the contrast. Once you are inside the house, you need no unusual ability to wander around and become familiar with it. What is hard and requires training is to understand written technical plans. This, I am sure, is how most people see mathematics. They are faced with a barrage of instructions, written in a language they cannot make sense of. It's not that they don't understand the mathematics, *they never get to it!* If they could only get beyond the plans and enter the house, they would find it as easy to wander around that mathematical house as an ordinary house.

I'm not saying that mathematicians find all mathematics easy. As in most things, some parts are harder than others. In a large house with many floors, it might take considerable effort to climb to the uppermost levels, and sometimes a door might stick or a room might be poorly lit. There are parts of mathematics that mathematicians themselves find difficult to understand. Still, unlike the majority of people, for whom mathematics seems a meaningless collection of apparently arbitrary rules for manipulating numbers, algebraic symbols, and equations, mathematicians go beyond the symbols and the equations and enter the house that those symbols and equations describe. Without the house, those rules, symbols, and equations are indeed meaningless, so it's hardly surprising that someone who cannot get into the house will find them incomprehensible. If you can enter the house, however, then everything seems real and has meaning. Thus, no matter how difficult it is to move around, doing so is a matter of interest and persistence, not the possession of some rare and magical talent or ability.

Of course, any metaphor has its limitations. My metaphor of the mathematical house underestimates the problems people often have with mathematics. Even if you cannot read a blueprint, provided you know the plans are for a house, you will at least have a rough idea of their purpose. But most people, faced with the numbers, symbols, equations, and rules for manipulating them, not only don't see what kind of object or world those numbers, symbols, equations, and rules describe, they don't grasp that they describe anything at all. Most people can't get beyond the symbols on the paper. It's like looking at a set of technical drawings and being unable to detect whether they depict a building, a machine, a spaceship, or a modern sculpture, or even whether they refer to anything at all. No wonder so many people find mathematics not just hard but completely devoid of meaning.

You might think that another limitation of the house metaphor is that, once you get inside a house, it takes no unusual ability to move around, whereas doing mathematics requires a particular ability that few possess. I don't believe this. Although I acknowledge that there are many levels of mathematical ability, I do not believe that a basic mathematical ability is any more unusual than an ability to talk. In the next chapter, I will argue that our brains are as equipped to move around the mental house of mathematics as our bodies are equipped to move around any physical house. I will also try to explain why so few people seem able to make use of this ability that I am claiming they possess.

In the meantime, for the remainder of this chapter, I'll try to give you a further glimpse of what it's like to live inside the mathematical house.

INSIDE THE HOUSE, OCCASIONALLY LOOKING OUT

For a mathematician, setting out to solve a mathematical problem or doing some mathematical research means first creating the mental "house": understanding the relevant mathematics and bringing it to the forefront of the mind. I am pretty sure I speak for all mathematicians when I say that this takes so much concentration that the mathematician must completely shut out the outside world. This period of intense concentration can last from a few minutes to days or weeks. (There's the Emily X parable again.)

As you might expect, faced with the pressures of daily living, most mathematicians learn to put their concentration on hold for brief periods. But as anyone who has lived with one of us can attest, when a mathematician is working intensely on a problem, she or he does not really engage the outside world. Rather, when it comes to everyday living, the mathematician functions on autopilot.

This focus on the mathematical world inside their own minds has resulted in numerous amusing stories of the absent-minded behavior of mathematicians. One of my favorites concerns Norbert Wiener, one of the foremost American mathematicians of the twentieth century. Wiener was at MIT at the time, and he and his family were moving to a larger house in the neighborhood. On the morning of the move, his wife went to great lengths to remind him that, when he left work that evening, he should go to the new house, not the old one. She handed him a slip of paper with his new address, realizing that by evening he would forget it. By evening, of course, Wiener was deeply engrossed in mathematical thought and absent-mindedly walked to his old home. As he headed up the garden path, he suddenly remembered that he was supposed to go to the new house. But what was the address? He began to search his pockets for the slip of paper his wife had pressed into his hands that morning. Where was it? There were dozens of slips of paper in his pockets, most of them covered with mathematics. Just then he noticed a child sitting on the porch of his old home. "Little girl," he said, "do you know where the people that used to live here have gone?" The girl looked up at him and smiled. "Yes, Daddy, of course I do. Mummy said you would go to the wrong home and would lose the address, so she sent me to fetch you."

This story is undoubtedly hugely exaggerated. No mathematician I have ever met would fail to recognize his or her own child! But heading for the wrong house or losing an address are typical of the kinds of things that mathematicians do when they are deeply engrossed in a problem. Indeed, an enterprising individual tracked down Wiener's daughter not long ago, and she had no recollection of her father ever not recognizing her, but she did attest to his frequent absent-mindedness. It goes with the territory.

I am sure that the level of concentration required to do mathematics affects many people's perceptions of their ability to do mathematics. Concentration at this level runs counter to everything for which the brain has evolved. A brain that makes itself oblivious to stimuli from the outside world may well not survive long enough for its owner to pass on his or her genes to the next generation. This is still true today; for our *Homo sapiens* ancestors, the risks were obviously far greater.

According to legend, the great ancient Greek mathematician Archimedes died as a result of being deeply engrossed in a mathematical problem. As the story is usually told, Archimedes did not hear the Roman centurion who had entered his house and who ordered him to stop working and identify himself. In frustration, the soldier ran him through with his sword. This level of concentration shows that mathematical thought requires considerable determination and effort—more than almost any other mental pursuit.

As I just remarked, I am sure that this is a factor that prevents many people from becoming proficient at mathematics. It isn't that they are incapable of intense concentration. Rather, they don't appreciate in advance the degree of concentration required. Hence, instead of giving it that concentration, they assume they just don't have the math gene. I am not claiming that other pursuits do not require intense concentration. But by and large, they require a high level of concentration *to do them well.* Mathematics requires intense concentration *in order to do it at all.* Without the concentration, the brain does not construct the symbolic house. And without the symbolic framework, the best anyone can hope for is to learn to perform various manipulations of *linguistic* symbols—the marks on the paper. The result is the all-too-familiar impression that mathematics is a collection of seemingly arbitrary rules to be applied in an uncreative, essentially mindless fashion.

To return to my main theme, how does it feel to a mathematician when she or he is working on a problem? By "a problem" I certainly do not mean an arithmetical problem that can be solved by following an established computational procedure. For instance, adding a long list of numbers is not a mathematical problem, it's a computational task. Broadly speaking, any problem you can solve with a calculator does not count as a mathematical problem.

The problem may nevertheless be a computational one, and the answer may be a number or an algebraic formula. But to be a mathematical problem, its solution must require some original thought.

Besides computational problems, there are problems that involve establishing some sort of logical relationship between various mathematical objects. Still other kinds of problem require the mathematician to formulate some new hypothesis and then prove it.

Here is an example of a problem that all mathematicians agree is definitely a *mathematical* problem.

HERE'S LOOKING AT EUCLID

The ancient Greeks made the seemingly mundane observation that some numbers can be evenly divided by smaller numbers whereas other numbers do not have this property. For example, 6 can be divided evenly by 2 and 3, but 7 cannot be divided by any number (apart from 1, which doesn't count since 1 divides into any number, and 7 itself, which again does not count, since any number can in a sense be divided by itself). Numbers that cannot be evenly divided by any other number are called *prime* numbers.

For example, 2, 3, 5, 7, 11, 13, 17, and 19 are the prime numbers below 20. All the remaining numbers below 20, namely 4, 6, 8, 9, 10, 12, 14, 15, 16, and 18, are *composite* (i.e., non-prime) numbers.

On the face of it, there is no reason to suppose that the distinction between prime numbers and composite numbers has any significance. But prime numbers turn out to be extremely important, and the more facts mathematicians can discover about prime numbers, the greater the number of consequences that follow. Part of the reason why prime numbers are so important is that, among the whole numbers, the primes are like the chemist's atoms—the building blocks from which all the others are built. To be precise, any composite number is a product of prime numbers; for example:

$$24 = 2 \times 2 \times 2 \times 3$$

Although not particularly difficult to prove, this result, that any composite number is a product of prime numbers, is so important in mathematics that it is given the impressive name "the Fundamental Theorem of Arithmetic."

An obvious question about prime numbers is: how many are there? As we just saw, of the first twenty numbers almost half are prime numbers. But the higher you go, the more the prime numbers seem to thin out. The question the ancient Greeks asked themselves was: do the primes eventually peter out?

The answer, given in Euclid's *Elements,* is to this day regarded as one of the most brilliant pieces of mathematics ever produced. Before I give that solution, I should point out that, strictly speaking, all the question requires is a yes or no answer. But that is not what mathematics is about. Mathematicians want to know why a statement is true or false. And the reason has to be in the form of a *proof,* a piece of logically sound reasoning that establishes the truth or falsity of the assertion in question.

In *Elements,* Euclid proved that the prime numbers go on forever—they are infinite in number. It is not known whether the proof in *Elements* is due to Euclid himself. In addition to presenting his own work, Euclid compiled much of the known mathematics of the time. Nevertheless, the proof is generally referred to as if it were Euclid's own.

To show that the primes continue forever, Euclid clearly could not exhibit all prime numbers. If indeed there are an infinite number of them, any attempt to list them all would be doomed to fail from the start. So he had to work more indirectly. What he did was to show that there is no largest prime.

Here is how he did it. Suppose there were, in fact, a largest prime number. Call it P.

Now, says Euclid, multiply together all the prime numbers from 2 to P, inclusive. No, you don't have to actually perform the computation. How could you? You don't have an actual value for P. Rather, you simply let N denote the result of that computation, whatever it is. That is:

$$N = 2 \times 3 \times 5 \times 7 \times 11 \times \ldots \times P$$

Now look at the number N + 1. It is obviously larger than P. Euclid claimed that this number N + 1 is a prime number. If he is right, then this will show that there is no largest prime number. Why? Well, look back at what we just said. We started out by supposing (against our better judgment, perhaps) that there was in fact a *largest* prime number. We decided to call it P. Now, according to Euclid, we have found an even bigger prime number, N + 1. Even larger than the largest one? Come off it! This is a logically inconsistent situation. Since we arrived at this state of affairs by supposing that there was a largest prime number, that must be the source of the inconsistency. (The only other things we did were to give that largest prime a name, P, and then specify how the number N was obtained by a straightforward multiplication, and neither of those steps could give rise to an inconsistency.) The conclusion is, therefore, that there is in fact no largest prime number.

So how did Euclid show that the number N + 1 is a prime number? This is the step we missed out above. He begins by asking what happens if N + 1 were not a prime number. In that case, by the Fundamental Theorem of Arithmetic, N + 1 must be a product of prime numbers. In particular, N + 1 can be divided by some smaller prime number, say M. But M divides evenly into N (because N is the product of *all* the prime numbers). Hence, when you try to divide M into N + 1 you are left with a remainder of 1. Once again we have an inconsistent state of affairs: M divides evenly into N + 1 and M leaves a remainder of 1 when you divide it into N + 1. And once again, this means that our initial supposition must have been false—in this case the supposition that N + 1 was not prime. So N + 1 is indeed a prime number.

That's Euclid's proof. Mathematicians regard it as one of the best examples ever of a neat, concise, and *elegant* proof.

Elegant? If you are a mathematician, it is hard to imagine that anyone cannot see the logical elegance in Euclid's proof. Moreover, the proof is so well constructed that it is hard for a mathematician to understand how anyone could fail to follow each step of the argument and be convinced by its overwhelming logic. No court of law has ever encountered a legal argument as logically sound and compelling as Euclid's 2,000-year-old proof of the infinitude of the prime numbers. And yet I know from teaching mathematics to hundreds of students that a great many not only fail to see how Euclid's proof could be described as "elegant" but are not able to follow the argument, and hence are unconvinced by it.

All of my students who have claimed not to understand Euclid's proof have demonstrated to me in many other ways that they are intelligent people who are quite capable of following logical arguments about their college courses, the advantages or disadvantages of various governmental policies, or how to invest their savings. I can only conclude that their inability to follow Euclid's argument stems purely from the fact that it concerns a completely abstract, symbolic world. In other words, it's the phenomenon we observed with the Wason test. If it were possible to reformulate Euclid's argument in terms of some familiar, everyday situation, I have no doubt that practically everyone would be able to follow it.

Indeed, if I were to provide such a translation of Euclid's proof, I am sure everyone would say it was "obvious." Unfortunately, because it talks about infinite collections and involves multiplication, Euclid's proof does not have such an equivalent, real-world reformulation. Consequently, if you don't "get it" in its mathematical form, then you won't get it at all.

Note that although Euclid's proof is about whole numbers, it's not computational. True, the very concept of a prime number involves division. Moreover, the proof involves multiplication and addition. But you never have to perform any actual computation. Both the question (Are there infinitely many prime numbers?) and Euclid's answer are about numerical *concepts,* not actual numbers. This is typical of most modern mathematics. Among all scientists, mathematicians are probably the worst at arithmetic. The reason is that, uniquely among scientists, mathematicians hardly ever have to perform an actual computation. It's not that they use computers to do their arithmetic but that numerical computations almost never arise in contemporary mathematics. Modern mathematics is about abstract patterns, abstract structures, and abstract relationships.

BACK INSIDE THE HOUSE

Faced with a problem to solve, the mathematician's first step is to bring into consciousness all the relevant mathematics—to create the appropriate symbolic house in the mind. Once the relevant mathematics has been completely understood and is uppermost in the mind—a process that, as we have observed, requires intense concentration—the mathematician starts to look for a solution. Exactly what this process involves depends on the problem. I'll use Euclid's proof as an illustration. It is by no means a typical example, for the simple reason that there is no typical example. But at least it provides some idea of what is involved in one particular case.

Imagine you are Euclid, trying to discover whether there are infinitely many prime numbers. First, you have to decide what the answer is likely to be: infinitely many or only finitely many. Maybe you can find some evidence that leads you to lean one way or the other. Otherwise, you might have to keep trying both ways in turn until one of them proves right. In the case of the prime numbers, Euclid probably had no evidence to help him either way. Maybe he tried both ways, or maybe he started with a lucky first guess that there are infinitely many prime numbers.

Suppose now that, for whatever reason, you (Euclid) think that there are infinitely many prime numbers. How do you set about proving it? Since you can't really cope with infinity, a sensible first step is to rephrase the goal: namely, you want to prove that there is no largest prime number. (This new formulation, although entirely equivalent to the original, does not refer to infinity explicitly.) Toward that end, a good approach is to investigate what would happen if there were, in fact, a largest prime number. (Trial lawyers do this kind of thing all the time. "Members of the jury, I would like you to ask yourself what would have happened if, contrary to what the prosecution has alleged, my client was not alone in the house at the time in question?") Now you have gone from talk about a possible infinity of prime numbers to looking at a single number, a supposed largest prime. Of course, you don't know what number it is. Indeed, you are hoping to prove it doesn't exist. But for the time being, you are supposing there is a largest prime number, and you give it a name: P.

Deep in your heart, you are convinced there is no such number P. So your next task is to see if the existence of P would lead to an inconsistent (and hence impossible) state of affairs. If you can do that, you will have proved your result.

What kind of consequence of the existence of P might involve an inconsistency? You have little to choose from. All you know about P is that it is the largest prime number. So your only possibility for an inconsistency is to show that there is a *larger* prime number. How do you do that?

So far, every step has been methodical. Many mathematical proofs proceed in just such a fashion. It is the next step that requires mathematical creativity. Who knows what thought processes he went through first, but what Euclid did next is the mark of the truly creative mathematician. Euclid says, take the number

$$N = 2 \times 3 \times 5 \times 7 \times 11 \times \ldots \times P$$

which you get by multiplying together all the primes, and then add 1 to it. With that single step, he has his theorem in the bag. Any mathematician will find the remainder of the argument straightforward. Since all of the primes 2, 3, 5, . . . , P leave a remainder of 1 when you divide them into N + 1, the number N + 1 must be a prime number, bigger than P.

As I mentioned, there is no such thing as a typical proof. Euclid's proof, however, shows how a proof can involve a number of routine steps that any competent mathematician could carry out, coupled with one or more truly creative and original steps that generally come only after the mathematician has thought about the problem for a very long time.

I have no idea how long Euclid thought about the infinitude of the primes before he came up with the key step in his proof. It could have been several days or maybe months.

During that time, Euclid may well have thought about other problems. Once the initial, high-concentration, "house-building" stage is completed, a mathematician can allow the concentration to ease from time to time. After a problem is firmly embedded in the mind, the subconscious often takes over, working on the problem without the mathematician's awareness. Many mathematicians have some of the best ideas while doing something else, perhaps as mundane as taking a bath or riding a bicycle.

The famous British mathematician J. E. Littlewood, one of the pioneers of modern number theory, once wrote (*The Mathematicians' Art of Work*, p. 114):

> I recently had an odd and vivid experience. I had been struggling for two months to prove a result I was pretty sure was true. When I was walking up a Swiss mountain, fully occupied by the effort, a very odd device

> emerged—so odd that, though it worked, I could not grasp the resulting proof as a whole.

Another famous British mathematician, Sir Roger Penrose, makes a similar observation in his book *The Emperor's New Mind* (pp. 418–25), adding the intriguing suggestion that aesthetics may play a role:

> My impression is that the strong conviction of the *validity* of a flash of inspiration (not 100 percent reliable, I should add, but at least far more reliable than just chance) is very closely bound up with its aesthetic qualities. A beautiful idea has a much greater chance of being a correct idea than an ugly one. At least, that has been my own experience. (Emphasis in the original)

Beauty in mathematics? Where did that come from? It's even more outrageous than elegance! Non-mathematicians often have great difficulty appreciating that there is such a thing as "beautiful mathematics."

Mathematical beauty is a very abstract beauty. Like fine music and great poetry, it is perceived not by the senses but by the mind. It is a beauty that could be perceived only by a brain capable of abstract, symbolic thought. The great English philosopher Bertrand Russell attempted to convey some sense of mathematical beauty by likening it to sculpture (*The Study of Mathematics*, p. 73):

> Mathematics, rightly viewed, possesses not only truth, but supreme beauty—a beauty cold and austere, like that of sculpture...sublimely pure, and capable of a stern perfection such as only the greatest art can show.

The mathematician J. L. Synge likened mathematics to nature (*Kandelman's Krim*, p. 101):

> The northern ocean is beautiful, and beautiful the delicate intricacy of the snowflake before it melts and perishes, but such beauties are as nothing to

him who delights in numbers, spurning alike the wild irrationality of life and baffling complexity of nature's laws.

Just as it requires training to appreciate the beauty in some forms of art and music, so too you have to "get inside" mathematics in order to appreciate much of its beauty. As Paul Erdös, a famous Hungarian mathematician who died in 1996, once remarked (quoted in Hoffman, "The Man Who Loves Only Numbers"):

It's like asking why Beethoven's Ninth Symphony is beautiful. If you don't see why, someone can't tell you. I *know* numbers are beautiful. If they aren't beautiful, nothing is.

Obviously, when mathematicians talk of the beauty in mathematics, they are not thinking about arithmetic. (In particular, when Erdös says he "*knows* numbers are beautiful" he is referring to the intricate patterns they weave, many of which he discovered.) Rather, mathematicians see beauty in the kinds of pattern we looked at in the previous chapter and in the way the same abstract pattern may appear in different guises.

Proofs too can be beautiful. Much as we speak of the beauty of a novel on many levels, the beauty of a mathematical proof lies in its logical structure, in the way the ideas are put together, and sometimes in the degree of surprise in a key step. Erdös used to refer to "God's book," a hypothetical volume that contained the perfect, most beautiful proofs of all the theorems of mathematics. Many times when a mathematician proves a result, depending on the nature of the reasoning used, colleagues will refer to the proof as "awkward," "cumbersome," "inelegant," "ugly," "neat," "elegant," or "beautiful." Occasionally, a proof will be so good that every mathematician agrees that it could not be improved upon. In that case, according to Erdös, the proof is "from the book," copied directly from the proof God Himself wrote. (Erdös once pointed out that you don't have to believe in God to be a mathematician, but you do have to believe in "the book.") Euclid's proof of the infinitude of the primes is an example of a "proof straight from the book."

As a beauty existing only in a symbolic world, mathematical beauty is both of, and reflective of, the human mind. Consequently, it is a great pity that most of this beauty remains hidden from general view, accessible only to those who achieve sufficient mastery of the field.

INVENTION OR DISCOVERY?

One question often asked about mathematics is whether it is invented or discovered. In my own experience, doing mathematics certainly *feels* like discovery. When I work on a mathematical problem or try to produce a mathematical proof, my sense is that the solution or the proof is "out there," waiting for me to find it. Yes, Euclid's proof of the infinitude of the primes involved an unmistakable element of human *creativity*. But it is a creativity of discovery, not invention. For instance, if Euclid had not found that proof, someone else almost certainly would have. Indeed, it is common in mathematics for several people, working independently, to simultaneously find essentially the same proof of a new result. Thus, the creative element in mathematics is not the same as the creativity involved in, say, playwriting. If Shakespeare had not lived, no one else would have written *Hamlet*.

But if mathematics is a process of discovery, it is discovery of a very odd sort. It discovers facts about an abstract world that is entirely the creation (!) of the human mind—an abstract world that arguably did not exist at all 5,000 years ago, and much of which is only a few hundred years old (some parts far younger than that).

Here we have another puzzle. If mathematics is a creation of the mind, whose mind creates (or created) it? Presumably, we each create our own mathematical world, inside our heads. How then can I be sure that the mathematics created in my mind is the same as that in yours? Are there many different ones?

The answer is that there is only one mathematical world. Whereas the conscious, directed creativity required to formulate a new mathematical concept or proof occurs in a single mind, the structure of the mathematical world itself is determined by the general structure of the human brain, and is thus the same for all our brains. The mathematical world is a product of the way the human mind encounters the physical world. Thus, mathematics is determined by both the world around us and by the structure of our brains.

The counting numbers, for example, do not exist for any of our fellow creatures. Nor did they exist for human beings until around 8,000 years ago, maybe less. Numbers are very much a creation of the human mind. They are an abstraction from the human ability to enumerate objects, one after another. Yet once you have a creature that can enumerate, the counting numbers can only turn out one way, starting with a unit (i.e., 1), and then increasing in size in single steps of that unit (2, 3, 4, etc.). Of course, different minds might adopt different linguistic signs and sounds to denote those numbers. But the numbers themselves will be the same, an automatic consequence of the ability to enumerate.

For another example, take the concept of a group, introduced in the last chapter. Recall that to have a group you must have: (a) some collection of objects; (b) a way of combining any pair of those objects to give another object in the collection; and (c) that way of combining objects must satisfy the three conditions listed on page 103. In some ways, the original definition of the group concept was a creative act. Nineteenth-century mathematicians decided that this concept was a useful and interesting one to study. In that sense, they invented groups. But the invention amounted solely to deciding what features of various real-world and mathematical phenomena to include in the definition. (Recall that they decided not to include a commutative law, i.e., they did not include axiom G4 on page 108.) Once that decision was made, the abstract world of groups was entirely determined. Creativity might be required to make discoveries within it, but there was no freedom to make that world different from the way it was. There were examples of groups in the world long before groups were defined. The mathematician simply abstracts from what is already there.

Because it is abstracted from the world around us, mathematics reflects that world, as well as the mental structure of the creatures that do the abstracting, namely ourselves. Thus, when I say that mathematics is a process of discovery, I mean it is a process of discovery both of the world around us and of ourselves as thinking beings living in that world.

How did our ancestors acquire the ability to create this abstract world of discovery?

The answer I shall give is closely bound up with our acquisition of language. Unfortunately, there is considerable confusion and misunderstanding about the nature of language. As a result, before I can show how acquisition of the language faculty prepared the way for the subsequent development of mathematics—something that many people believe (erroneously) they know little about—I must first take a hard look at language, which everyone feels (erroneously) they understand just fine. In particular, what exactly is language, how does it differ from a system of communication, and how does practically every one of us manage to achieve fluency in at least one language at a very early age?

·6·

BORN TO SPEAK

DID SHAKESPEARE KNOW THE DIFFERENCE?

Fie, fie upon her!
There's language in her eye, her cheek, her lip,
Nay, her foot speaks; her wanton spirits look out
At every joint and motive of her body.

THE STYLE is unmistakably Shakespeare's. The quotation is taken from his play *Troilus and Cressida,* Act IV, Scene V. As usual, the Bard manages to find just the right words to convey his meaning. And as usual, some of these words must be understood metaphorically. Cressida's foot does not literally "speak," nor do her wanton spirits actually "look out." But did Shakespeare realize that his use of the word "language" was also metaphorical? It depends on what he understood by the word "language." For today's language scholar, however, there is no question: what can be found in "her eye, her cheek, her lip" is not language but communication.

Aren't language and communication the same thing? Isn't language just a mechanism for communication? If you think the answer to either question is yes, you are certainly not alone. Many people confuse the two. I shall begin this chapter by explaining why language and communication are not at all the same thing.

It would of course be absurd for me to claim that we cannot communicate using language. After all, communicating my ideas to you through language is precisely why I wrote this book! Language is indeed a mechanism for communication. But it is not *just* that. Moreover, in Chapter 8, I will argue that language did not develop primarily as a communicative device. To achieve that, I will have to put in considerable effort, because many authorities claim exactly the opposite.

On the other hand, I suspect I will have a far easier time persuading you that language and communication are not the same. Indeed, people mostly confuse the two only when they have not had occasion to stop and reflect on the difference between an activity and the means of carrying it out. For example, an automobile is a device for driving, but an automobile is not the same as driving. One is the activity, the other the means. Similarly, a pen is a device for writing, but pens and writing are not the same thing. So too for language and communication. Communication is something we do with language. Language is a medium with which we can communicate. But they are not the same thing.

The distinction between the tool and the use to which it is put lies beneath my claim that Shakespeare's use of the word "language" in the epigraph was metaphorical for "communication." But I had better not leave it there, since to do so would leave me open to a counter-attack by anyone who understands human behavior. Here is how that counter-attack would go:

"When someone communicates using an eye, a cheek, a lip, a foot, even wanton spirits, they are indeed using a 'language'—what we call *body language*. Body language may not be the same as English, French, or Chinese, but it is still a language. It has the equivalent of words—a raised eyebrow, a frown, a shrug of the shoulders, a 'wanton look,' and so forth—and there are rules for how and when to use those gestures, just as there are rules for how to use the words of English, French, or Chinese."

Perhaps you have been nodding in agreement. So instead of developing it further, let me say why it is wrong. It is wrong because what it describes—"body language"—is not a language. It's a *system of communication.* For many purposes, it is a very good system of communication. What is more, our evolutionary ancestors almost certainly used a system of gestures and grunts to communicate with one another long before language appeared on the scene.

But to a linguist, body language is not language. Nor indeed are the communication systems—many of them quite sophisticated, elaborate, and expressive—used by all our fellow creatures on the earth.

So what do the experts mean by language? And how does language differ from the communication systems of apes, monkeys, bees, dogs, and dolphins? Let me start by giving a brief summary of my argument.

Communication systems are generally mixtures of sounds, gestures, facial expressions, skin coloration, and bodily movements that enable creatures to inform fellow creatures of their current emotions, current needs, current desires, pending action, impending danger, and the location of food supplies. They allow the conveyance of single ideas such as "Danger!," "Leopard!," "I'm hungry," "Object A has property B," or "X is over there." They do not, however, permit more complex messages. Moreover, they can generally convey information only about the immediate environment; indeed, much of their information-carrying capacity depends on that environment. (There are some exceptions, where a particular signaling system can convey information about some other location or a remote object, such as the well-known "waggle dance" that bees use to inform their fellow colony members of a source of food.)

A language, by contrast, has a combinatorial structure (what linguists call syntax or grammar) that allows the expression of far more complicated ideas. The rules of syntax tell you how to put words together to form meaningful expressions that convey complex ideas. Because of this structure, language may convey ideas about circumstances and events outside the immediate environment.

One obvious distinguishing feature of any human language is the presence of words that do not refer to anything in the physical world. Whereas the signs and signals of animal communication systems all refer to something specific in the world—such as an object, a location, an action, or an emotional state—languages have a great many signs whose purpose is within language itself. English words such as *if, then, because, unless, and, or,* and *every* have a syntactic function: their role is to provide the means of combining simple ideas into more complicated expressions.

In simplistic terms, you get a language when you add syntax to vocabulary. Part of the reason this description is simplistic is that it gives no hint of the enormous difference "adding syntax" makes. As we shall see, the addition of syntactic structure does more than extend the expressive power of the system. It changes the system into something very different, which can be used in very different circumstances, for quite different purposes.

In this chapter and the two that follow, I will try to convey some sense of the huge difference between a language and a system of communication. But for an initial (and fairly crude) sense of the magnitude of this difference, let's see how many sentences of at most twenty English words can, in principle, be produced using English syntax.

First of all, as languages go, English has a fairly large vocabulary. A good English dictionary will list up to 450,000 words, but this is far more than the number actually known to a typical English-speaker. A bright six-year-old American will know between 13,000 and 15,000 words. The average American adult has a vocabulary of between 60,000 and 100,000, the avid reader considerably more. But these figures apply to words an individual can recognize and understand in context. The vocabulary we actually use on a day-to-day basis is considerably smaller. For example, Shakespeare used around 15,000 words in all of his works combined. Let's assume you have a speaking vocabulary as large as Shakespeare's 15,000 words. How many sentences of at most 20 words can you, in principle, produce?

Statistical studies of language have shown that if a native English-speaker is interrupted at a random point while uttering a sentence in English, there are, on average, about ten possible words that could come next and allow for the utterance to be continued to produce a grammatical sentence. At some points there may be only one possible next word; at other points there are thousands of choices. The average is about ten. This means that there are of the order of 15×10^{22} (that is, 15 followed by twenty-two zeros) grammatical English sentences of twenty words or less. (To arrive at this figure, take 15,000 possible choices for the first word, multiply by 10 for the second word, 10 for the third, and so on.) At the rate of five seconds per sentence, it would take you 150,000 trillion years to say them all out loud, with no breaks for eating or sleeping.

That's a phenomenal number of sentences. The number of grammatical sentences of twenty words or less that you can understand will be considerably larger, about 6×10^{23}, given your much greater recognition vocabulary. How can your brain handle such a vast number of possibilities? Clearly, not by storing them all in your memory until they can be brought into consciousness when the occasion presents itself. Rather, your brain must have some mechanism for generating—and analyzing—grammatical sentences on the fly.

Whatever that mechanism looks like, it operates unconsciously and with remarkable efficiency. For example:

Rearrange these ten English words to give another grammatical sentence.

After at most a few seconds, you might find:

To give another grammatical sentence, rearrange these ten English words.

With further thought you may find other permissible rearrangements, although many of them won't make much sense. But there are 3,628,800 possible arrangements of those ten words, and the vast majority of them are not grammatical sentences. You certainly did not sift through them at random looking for the few grammatically correct ones. If you'd gone about it that way, you probably would not have found even one alternative arrangement. Somehow, you knew the kind of rearrangements that *might* give a grammatical sentence and you checked *them*.

Notice that the only difficult part of that last exercise was generating the alternative sentences. It takes no effort at all to decide if a given sequence of words is a grammatical sentence. The mechanism for assembling words to form grammatically correct sentences is such a fundamental part of our brains, and so precise, that—apart from an occasional hesitation or disagreement—when presented with a sequence of up to a dozen or so words, all native speakers of a particular language will agree—*immediately, instinctively, and without conscious reflection*—whether that sequence is grammatically correct.

How do we do this? It turns out that it's not really a matter of putting words "in the right order." Rather, the ordering of the words in a grammatical sentence is just a surface manifestation of a deeper structure. That deeper structure lies at the heart of this thing called language.

VARIATIONS ON A SINGLE THEME

There are approximately 5,000 different languages in use today—an exact count can't be given because linguists sometimes disagree as to whether two languages are really different or whether one is a dialect of the other.

Before linguistics arose at the end of the nineteenth century, studies of language mainly consisted of historical accounts of how particular languages developed—English, French, Rumanian, or their much earlier precursors, Proto-Germanic, Proto-Indo-European, and so on—and of the relationships between them.

Starting with the work of the Swiss scholar Mongin-Ferdinand de

Saussure in the early twentieth century, however, linguists have examined language as a phenomenon, a human capacity. Their aim is to develop a science of the structure of language in much the same way that physicists seek a science of the structure of the universe.

One of the most significant discoveries the linguists made was that, despite their different vocabularies and different word orders, all human languages share an underlying structure—they are all variations on a single theme. One of the first linguists to discover this fact was an American, Noam Chomsky, in the 1950s and 60s.

Chomsky began by observing that grammaticality—whatever it is—is not the same as making sense. For example:

Colorless green ideas sleep furiously.

Any native English-speaker will agree at once that this is a grammatically correct sentence, and yet it makes absolutely no sense. Every adjacent pair of words is an oxymoron.

Since grammaticality does not require sense, Chomsky concluded, it must be structural. It might therefore be possible to formulate a collection of rules that specify how words must be put together to produce grammatically correct sentences. Chomsky called such a collection of rules a grammar, thereby giving precise, scientific meaning to a word that had hitherto been restricted to instruction on language etiquette, where it was invariably coupled with the words "good" and "bad."

An example of such a grammatical (or syntactic) rule is that a singular noun phrase followed by a singular verb phrase gives a sentence. For example, take the singular noun phrase, *The big dog,* and follow it with the singular verb phrase, *ran into the street,* and you get the sentence *The big dog ran into the street.* In his ground-breaking book *Syntactic Structures,* published in 1957, Chomsky presented a large number of rules of this kind, expressed in an algebra-like notation. For instance, the rule we just saw is:

$$NP_s\ VP_s \rightarrow S$$

This says that a singular (the first subscript s) noun phrase (NP) followed by a singular (the second subscript s) verb phrase (VP) gives (→) a sentence (S). Further rules describe how to construct noun phrases and verb phrases. For example, the rule

$$DET_s\ A\ N_s \rightarrow NP_s$$

says that a singular determiner (such as *a* or *the*) followed by an adjective, in turn followed by a singular noun, gives a singular noun phrase, such as *The big dog.*

Since these rules describe how to generate grammatically correct sentences, Chomsky referred to any collection of such rules as a generative grammar.

Although generative grammars turned out to be influential in the design and implementation of higher-level computer programming languages, they were soon all but abandoned by the linguistics community. (Although not before they had transformed linguistics from a narrow and decidedly "soft" social science to a broad discipline that uses methods of both social science and mathematics.) A number of problems led to their loss of favor. For one thing, it takes a large and unwieldy collection of rules to generate the wide range of grammatical sentences that arise in everyday life. Some rules contradict each other, requiring additional rules to legislate between them.

Even with a fairly large collection of rules, many of them providing sophisticated mechanisms to ensure agreement in case (singular or plural, first, second, or third person) and tense, generative grammars tend to produce stilted, machine-like sentences that resemble the artificial-sounding utterances of robots in science-fiction movies. To address this problem, Chomsky postulated the notion of "deep structure." Here, briefly, is the idea.

First, let's observe that there are several different kinds of sentence.

Sentences that state a fact are called declarative sentences. Those that ask a question are called interrogatives. Sentences that attempt to get the hearer to perform some action—such as commands—are called directives. Sentences that express a psychological state, such as apologizing or praising, are called expressives. Finally, there are sentences that, when uttered by a suitably qualified person, bring about their content simply by virtue of being uttered, such as a president or prime minister declaring war or a minister declaring a couple man and wife. These sentences are called performatives.

Any sentence has a deep structure, which depends on its type. For instance, the purpose of a declarative sentence is to say who did what to whom, how, where, when, and maybe why. More precisely, the sentence identifies some or all of the following: an agent, an action, an object, a means, a location, a tense, and a reason. The sentence's deep structure is the allocation of particular words or phrases to the various categories of agent, action, object, etc. The order of these components may vary, giving different "surface structures," but the deep structure remains the same. For example, each of the following provides the same information:

The major killed the maid in the library with a dagger.

The major killed the maid with a dagger in the library.

The maid was killed by the major in the library with a dagger.

The maid was killed by the major with a dagger in the library.

Chomsky would say that all four sentences have the same *deep structure* but different *surface structures.*

Chomsky's idea was that generative grammars produce deep structure, and a further collection of rules, called transformation rules, turn the deep structure into various grammatical sentences by swapping word order, adding endings to words to produce agreement, and so forth.

Since we have no means of accessing the mental processes that produce language, we cannot know whether people actually produce sentences by first generating deep structure and then massaging it into a suitable grammatical sentence. But even as a theoretical device, the approach leaves much to be desired, and not just because of its complexity.

WORDS IN BOXES

Possibly the greatest objection to the generative/transformational grammar approach is that it gives too much prominence to word order. Grammatical rules of the kind "an X followed by a Y gives a Z" tell you how to string words and phrases together in a linear fashion, like beads on a string. But that is not really what sentences are about.

In science, the most difficult challenge is often to decide which are the truly important issues and which are superficial. In the case of language, linguists took a major step forward when they realized that word order is a comparatively superficial feature of language. For instance, my earlier example of the 10 English words, where only a few of the 3,628,800 possible orderings yield grammatical sentences, is surprising only if you think that creating a sentence is primarily about sequencing words in a linear fashion. Sequencing is only the last step in sentence formation, by which time the number of possible orderings is extremely small.

It took several decades to find the right way to approach syntax, but eventually linguists (Chomsky among them) started to figure it out. The key is to regard the phrase, not the word, as the basic unit for constructing sentences.

Sentences are themselves phrases, as are clauses. And any phrase is itself made up of smaller phrases, right down to the level of single words. (Linguists effectively regard single nouns and verbs as phrases in their own right.)

In this approach, a sentence is regarded not as a row of beads on a string but as a collection of nested boxes. Open up a sentence box and you will find various clause or phrase boxes. Open up a phrase box and you find further phrase boxes, and possibly some words. If you keep opening boxes, you will eventually come to boxes that contain just words but no other boxes.

For example, consider the sentence, shown in Figure 6.1, *Noam wrote that book*. Think of it as a phrase box labeled S (for sentence). When you open box S, you find it contains two smaller phrase boxes, one labeled N (for noun), the other labeled VP (for verb phrase). The box N contains just one object, the word *Noam*. But the box VP contains two further boxes, one labeled V (for verb), the other NP (for noun phrase). If you open the box V, you will find just one object, the word *wrote*. The box NP, however, contains two further boxes, one labeled DET (for determiner), the other N (for noun). The box DET contains the word *that;* the box N contains the word *book*. And now there are no more boxes to unpack.

Mathematicians call a diagram of the kind shown in Figure 6.1 a tree. Trees are an extremely convenient way to depict any kind of hierarchy. Family trees are a familiar example, as are the organizational charts (or

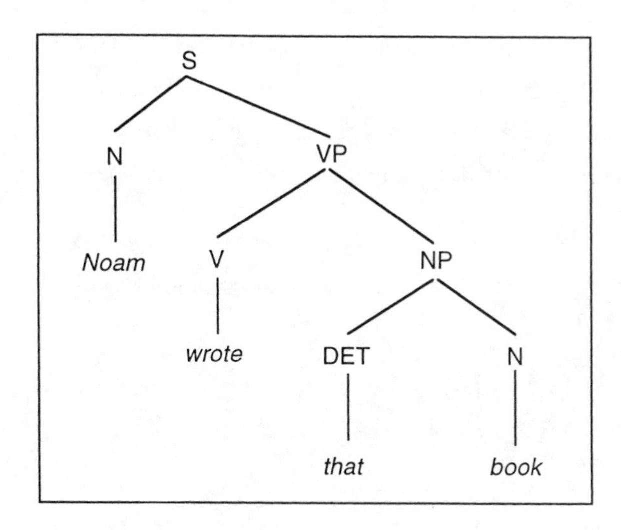

FIGURE 6.1 A Simple Parse Tree

reporting trees) of large companies. A tree showing how a sentence or some other linguistic unit splits up, as in Figure 6.1, is called a parse tree. The boxes labeled S, N, VP, V, DET, and N (the "nodes" of the tree) are all phrases. Thus the parse tree shows exactly how each phrase is built up from (or parses into) simpler phrases.

A grammar for the language is a collection of rules for putting various boxes into other, "larger" boxes. For instance, the generative grammar rule $NP_S\ VP_S \rightarrow S$ (in which ordering matters) corresponds to a rule that says: if you take a singular NP box and a singular VP box, you can put them together into an S box. This gives us the parse-tree structure shown in Figure 6.2.

Notice that, unlike Figure 6.1, which depicted the phrase structure of an actual sentence, Figure 6.2 represents a grammatical rule, so the ends of its branches contain no words. When the empty boxes labeled NP_S and VP_S are filled suitably, then you get a sentence.

An important point to remember about trees is that the lines depict the only relationships between the nodes. The horizontal arrangement of a tree is merely an artifact of the way it is depicted on the page. In order to make it easier to "read off" the structure of the tree and relate it to a given sentence, the tree is generally drawn to match the standard word order in the sentence. But this is a matter of presentation. The left-to-right orderings are not part of the tree's actual structure. (When you open a box and find it contains two or more smaller boxes, you can open those boxes in

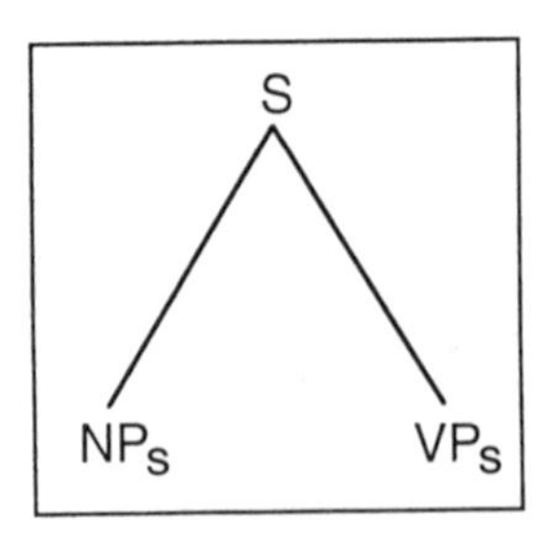

FIGURE 6.2 A Very Basic Parse Tree

whatever order you please.) Thus the two parse trees shown in Figure 6.3 are two ways of drawing *the same tree.*

This fact about trees reflects the linguist's realization that word order is just a "surface" feature of a particular language. The parse tree captures Chomsky's deep structure and ignores surface structure. Consequently, trees provide a more accurate way to express grammatical rules than do algebraic formulas. The single tree in Figure 6.2 corresponds to the two algebraic rules $NP_S\ VP_S \rightarrow S$ and $VP_S\ NP_S \rightarrow S$.

"Nested box" grammars, where the rules are represented by parse trees, are generally referred to as phrase structure grammars. Phrase structure grammar became particularly important with the discovery of a remarkable fact: studies of various languages, including the sign languages used by the deaf, show that when word order is ignored, they share essentially the same underlying grammatical structure. That is to say, apart from minor differences, they have the same phrase structure grammar. They all have nouns, verbs, modifiers (adjectives and adverbs), pronouns, and prepositions. They have the same kinds of phrases. And apart from differences in word order and a few other details, they all assemble smaller phrases into larger ones in essentially the same way.

Thus, the many different languages we find in the world are all cut from the same cloth. What distinguishes one from another—other than the different vocabularies (and in some cases the different alphabets for written language)—is that each language has its own set of rules that govern the

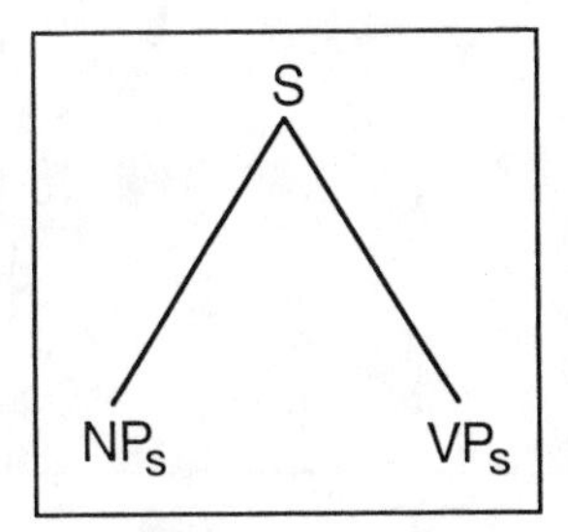

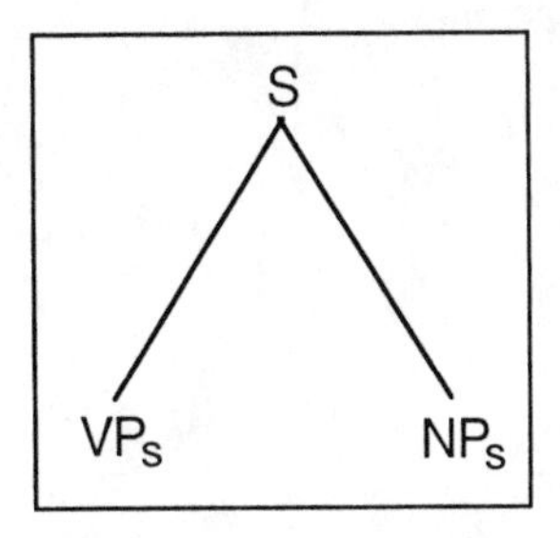

FIGURE 6.3 Two Depictions of the Same Parse Tree

word ordering within each phrase and the ordering of phrases within a sentence.

I should stress that what languages share is not some particular tree structure but a general way of putting phrases together according to tree structures. In any one language, the actual tree that you get from a particular sentence will vary according to the sentence. For example, Figure 6.1 gave the parse tree for *Noam wrote that book*. That is a particularly simple example, whose parse tree is binary, meaning that each node splits into exactly two "child" nodes (each phrase box has exactly two members) until you get down to individual words. Now consider the sentence *Noam gave that book to Susan*. This has the more complicated parse tree shown in Figure 6.4. In this tree, the VP node has three "children" (the VP box contains three boxes), a V-node, an NP-node, and a PP-node (prepositional phrase). The PP-node, in turn, has two "children," a P-node (preposition) leading to the word *to*, and an N-node leading to the word *Susan*.

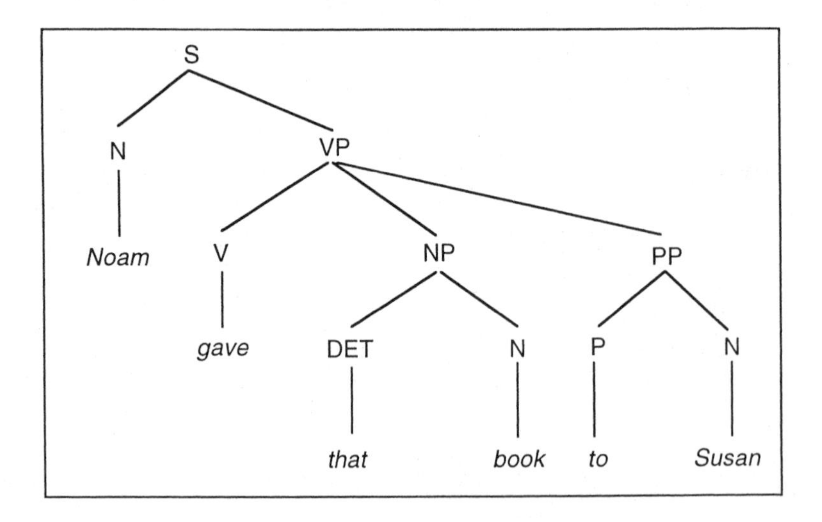

FIGURE 6.4 Another Parse Tree

In any language, a range of different tree structures can arise as parse trees. Thus, although phrase structure grammar helps us to see how all the world's languages share a common underlying grammar, we are still faced with considerable complexity. In order to understand how we produce, recognize, and understand grammatical sentences, we need to make sense of that complexity. We must look for the rules that determine what kinds of trees are allowed. This is where linguists made a startling observation.

THE KEY TO SYNTAX

When we look carefully at the tree structure of phrases, it turns out that all phrases—noun phrases, verb phrases, adjectival phrases, prepositional phrases, even sentences themselves—have the same basic containment structure, represented by what I shall call the fundamental language tree, shown in Figure 6.5.

The fundamental language tree is a bit like the basic Lego building block. Just as that one standard shape can be fitted together in different ways to make a wide variety of different structures, so too repeated application of the fundamental language tree produces all the tree structures of grammatical parts of speech.

We can view the fundamental language tree as a rule for putting together three phrase trees (or other linguistic units) to give one larger phrase tree. When three phrases are put together to give a phrase using the fundamental language tree, the principal constituent of the resulting phrase is called the head. It always goes in the position labeled HEAD in Figure 6.5. If the head is a noun, the resulting phrase is a noun phrase; if the head is a verb, the phrase is a verb phrase; and so on. The head is coupled with what is called the complement, going into the position labeled COMP in the diagram. Finally, the entire phrase has what is called a specifier, which goes in the position labeled SPEC. Essentially, the specifier and the complement provide informative packaging for the head.

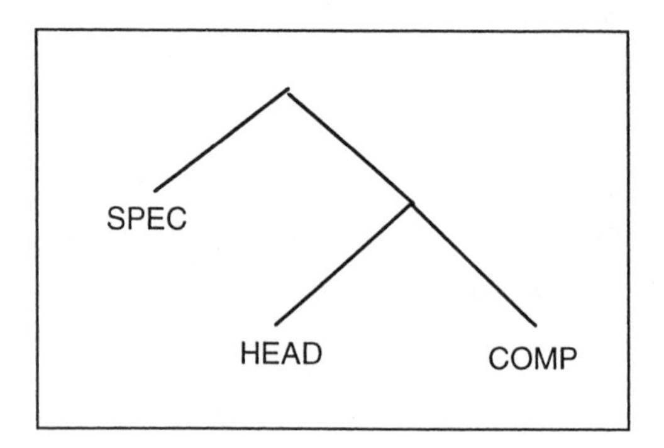

FIGURE 6.5 The Fundamental Language Tree

For example, Figure 6.6 illustrates how we can use the fundamental language tree to form the noun phrase *the red door.* The head of this noun phrase is the noun *door.* The complement to the word *door* is the adjective *red.* The specifier is the determiner *the.*

Now, it's probably not obvious from one simple example that this rule can generate all possible grammatical sentences in all human languages. Unfortunately, it would take us too far off our present course for me to show you how it works in other cases. (Remember, our goal is to understand how humans are able to do mathematics.) Instead, I'll give you just one more simple example here—another simple noun phrase—and direct you to the appendix for a whole range of examples. The appendix will also explain the significance of the terms "head," "complement," and "specifier."

FIGURE 6.6 Construction of the Noun Phrase *the red door*

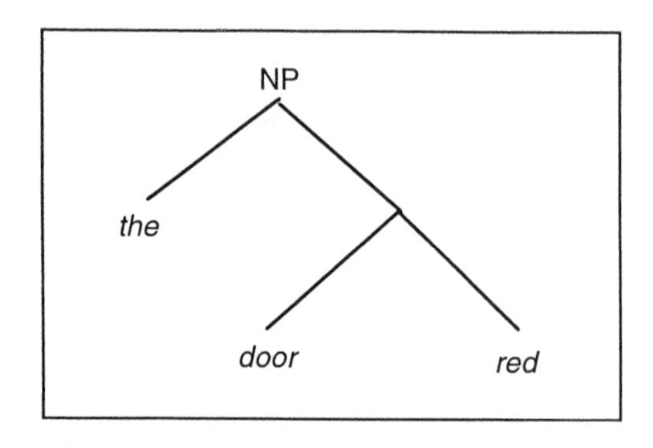

In the example *the red door*, we used the fundamental language tree to combine three words. But we can use the same rule to combine words and phrases. Figure 6.7 shows how we can generate the more complicated noun phrase *the house with the red door*. In this example, the word *house* is the head, the prepositional phrase *with the red door* is its complement, and the word *the* is the specifier.

The important point is that a *single, very simple* structure is enough to generate (by repeated application) all sentences. But why should we take the tree in Figure 6.5 as the fundamental building block? Surely, the two-pronged tree shown in Figure 6.2 (without the labels on the nodes) is more basic still, and the fundamental language tree can be built up from two copies of that simpler tree. But the point of this analysis is to identify what gives rise to the complexity of language. The tree in Figure 6.2 is too simple. If we take that as our basic building block, we will need to develop an elaborate list of rules for how such trees may be assembled. With the fundamental language tree, we don't need such a list. (Similarly, the definite shape of Lego blocks lets us connect them together to build a wide variety of structures. Plain rectangular blocks would be simpler, but they would not do the same job.)

In fact, the fundamental language tree captures the heart of the syntactic structure of all human languages so well that it is hard not to conclude

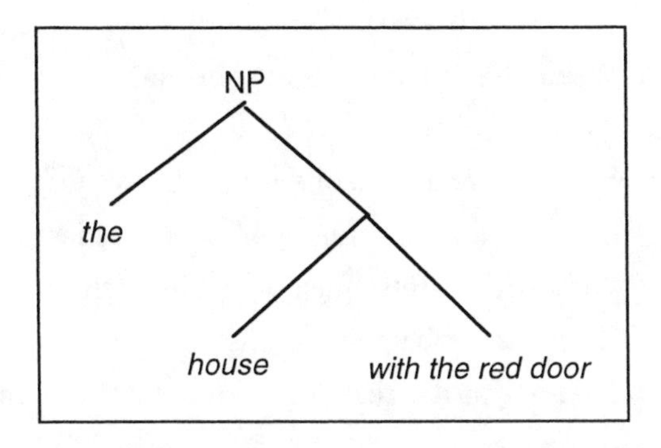

FIGURE 6.7 Construction of the Noun Phrase *the house with the red door*

that this must be how the human brain handles language—that the combinatory rule expressed by the fundamental language tree must be "hard-wired" into the brain.

This gives us a simple picture of the way we produce grammatical sentences. We generate *any* grammatical sentence by the repeated application of this one rule for "putting things together," coupled with rules for the order in which certain things go (e.g., subject before verb before object) and rules for agreement (e.g., singular/plural, gender, and tense). Whereas the rules for order and agreement vary from language to language, and the actual parse tree that you get will vary from sentence to sentence, the basic "putting things together" rule—the fundamental language tree—is universal. For this reason, the grammar determined by the fundamental language tree, the core grammatical structure of all human languages, is sometimes called universal grammar.

To a professional linguist, the fundamental language tree seems pretty neat. It's clean, lean, and precise, and provides a single mechanism for generating all phrases, clauses, and sentences.

Well, not quite. To be honest, it's not as clean as it looks. There are a number of technical issues I did not mention.

One objection to the kind of simplified description I just provided is that, like all metaphors, the "how to put boxes into boxes" picture is misleading if taken too far. In particular, the contents of one box can affect the contents of the others that are packed with it into a larger box. For instance, tense, gender, and plurality are required to agree, a matter that different languages handle in various ways through word endings, vowel changes, pronouns, and so on.

Another question is, when someone utters (or writes) a sentence in a particular language, how do the actual words used reflect the underlying tree structure (which they surely must)? Much of the work is done by non-content words. Roughly half the words in a typical English sentence do not refer to anything in the real world—they are not content words. Rather they perform a grammatical function, internal to language. Their purpose, roughly speaking, is to help keep a record of the way the boxes are assembled, so that they may be correctly unpacked. Likewise, the various rules concerning agreement and word placement (what goes next to what, what must be close to what, etc.) also help us keep track of how the boxes are stacked.

Now, it is true that the functions performed by the various grammatical words, agreement rules, and placement rules can be explained in terms of the fundamental language tree. But by concentrating on the tree structure alone and ignoring all other aspects of linguistic structure, we run the risk of portraying syntax as far simpler than it really is. This caveat having been made, however, the account I have given does provide the major core of the theory. I would liken it to the (classical) molecular theory of matter. We know that this theory is a gross simplification of the nature of matter. Nevertheless, it has served as a foundation for chemistry and much of biology for a century now, and doubtless will continue to do so for many years.

HOW DO TWO-YEAR-OLDS MANAGE TO LEARN ALL THIS STUFF?

The fact that language is more complex than the fundamental language tree might suggest only adds weight to the following question. Since even my simplified presentation of the fundamental language tree takes considerable effort to understand, how is it that children master syntactic structure by the time they are three years old?

For one thing, we should be careful to distinguish an ability from a scientific analysis of that ability. Three-year-old children can also walk on two legs but they are not up to the (highly complicated) mathematical analysis of two-legged perambulation. Although it can require considerable effort to see how a particular phrase or sentence can be generated from the fundamental language tree (see the appendix), the entire theory is built around the repeated application of the *one very simple combinatorial rule* illustrated by the fundamental language tree. If the human brain somehow embodies that tree structure, then "learning" to speak and understand grammatical sentences need be no more puzzling than "learning" to walk.

If syntactic structure—in the form of the fundamental language tree—is "hard-wired" into our brains, then "learning to talk" need only involve learning the vocabulary of our particular native language, together with some information about word order. We then simply slot the appropriate words into the various locations in phrase structure trees to produce grammatical sentences.

This explains the two-stage language acquisition process we observe in all young children: a typical child will spend the first two years of its life acquiring an ever-larger stock of useful words, but all its utterances will comprise single-word or simple two-word subject–property combinations. Then, some time between its second and third birthdays, the child will suddenly start to produce fully formed sentences of near-perfect grammaticality. Here is what (probably) happens.

During the first stage, by listening to the adults around it, the child simply learns new words and picks up the prevalent word order patterns. Then, when it has learned enough about word order and its vocabulary is sufficiently large to be able to form meaningful sentences, the built-in syntactic structure (the fundamental language tree) swings into play to produce grammatical sentences. In other words, vocabulary and word order are learned; grammatical structure is innate in much the same way that spinning webs is hard-wired into a spider's brain.

Since the fundamental language tree is the same for all the world's languages, but word order and other surface features differ, this theory also explains how a child will acquire whatever language it is exposed to. For instance, an American child growing up in Japan will learn to speak fluent Japanese, even though Japanese and English are quite different. The word order is different, for one thing. In Japanese, the verb generally comes after the object, whereas in English the verb usually follows the subject.

(Thus, in English we say *Syun eats sushi,* but the Japanese equivalent translates word for word as *Syun sushi eats.*) A theory of language that considered word order part of the "hard-wired," basic grammar passed on through the genes would predict that an American child would adopt English word order wherever he or she grew up. But if the basic syntactic structure captured by the language tree is innate and genetic, whereas such surface features as word order are cultural and learned, then we can understand not only how young children suddenly begin to produce grammatical sentences but also why all the world's languages have the same universal grammar.

But what do we mean when we say that human brains are "hard-wired" with the fundamental language tree? Certainly not that neuroscientists will one day discover tree-like circuits in the brain. Rather, we mean that (a) we have an innate capacity to acquire a language, and (b) this innate capacity can be described in terms of a phrase structure grammar such as the one captured by the fundamental language tree.

Specific evidence that humans have an innate capacity for syntax is provided by detailed studies of language acquisition in children. I'll cite just one of many such studies, based on the process of turning a statement into a question.

Suppose, for the sake of argument, that children learned how to turn a statement into a question not by applying a built-in syntax capacity but by formulating a rule for moving words, based on how they hear adults making statements and asking questions.

Take the sentence:

The mailman is at the door.

This is most simply turned into a question by moving the auxiliary verb *is* to the start of the sentence, to give the question:

Is the mailman at the door?

Suppose a child learned how to make this transformation by formulating a simple word-movement rule. The most obvious—and therefore the most likely—rule would be: scan the sentence to find the auxiliary verb *is* and move it to the beginning of the sentence.

What would happen, then, if the child encountered the following sentence?

The woman who is going to take your photograph is at the door.

If the child simply scans the sentence until she finds the auxiliary *is,* and moves it to the start, the result is:

Is the woman who going to take your photograph is at the door?

The child has obviously picked up the wrong occurrence of *is.* In fact, children never do this. In the mid-1980s, the psychologists Stephen Crain and Mineharu Nakayama carried out extensive tests of three- to five-year-old children, which confirmed that they almost invariably perform the correct transformation. They do not simply pick the first occurrence of the auxiliary verb, but instead look for the appearance of the auxiliary in the main verb phrase of the sentence. To do this, the child has to be able to recognize such parts of speech as subject phrases, verb phrases, and object phrases, and to understand how these parts of speech fit together to form sentences.

In other words, although it would be absurd to say that young children understand the fundamental language tree (and the theory that supports it), they do have mastery of all the phrase structure that the fundamental language tree captures. Given this implicit understanding of sentence structure, the child can apply the correct rule—look for the occurrence of the auxiliary verb *is* in the main verb phrase of the sentence and move it to the start of the sentence—to produce the correct question:

Is the woman who is going to take your photograph at the door?

Incidentally, many years earlier Chomsky had observed that young children cannot acquire such a complex statement-into-question procedure by imitating a number of specific instances, and for a very simple reason: parents rarely use sentences like *Is the woman who is going to take your photograph at the door?* when speaking to their small children. Nor do they provide examples of statement-question pairs to illustrate the transformation rule.

Chomsky's observation leads to an inescapable conclusion. If young children do not learn syntax by imitating adults, they must be born with it. In other words, the deep-structure rules must be hard-wired into their brains.

Now we are ready to turn to the question I mentioned earlier: how did the human race acquire syntax in the first place, and for what purpose?

·7·

THE BRAIN THAT GREW AND LEARNED TO TALK

TWO PUZZLES ABOUT THE EVOLUTION OF LANGUAGE

THE MOST FAMILIAR evolutionary account of how the human brain acquired language is that the benefits of increasingly rich means of communication led from a fairly primitive vocal communication system, through systems of increasing complexity, until full-blown language emerged. According to this theory, the first step along the road to language was our early hominid ancestors' gradual acquisition of a richer and richer vocabulary (without syntax). As we shall see, there is evidence suggesting a steady growth in vocabulary over a period of up to three and a half million years. During this time, verbal communication would have consisted of nothing more complicated than primitive utterances along the lines of "Give!" or "Go!" and possibly some simple object-property utterances, such as "Me angry," or "Lion coming," very probably accompanied by gestures and other physical means of communication. Linguists refer to the linguistic part of this kind of crude communication system as a protolanguage. The grammatical rule (or protosyntax) of such a protolanguage is the two-pronged tree shown in Figure 7.1, which I shall call the protolanguage tree.

The protolanguage tree is the same tree as in Figure 6.2, but without labels on the nodes. The absence of labels indicates that we are to view the tree as representing a rule for putting things together. Presenting the rule as a tree indicates that the order in which the two things are put together is not important. It also provides a visual comparison with the fundamental language tree.

There is evidence that chimpanzees, monkeys, and apes can be taught protolanguage. (However, I should note that research on animal language acquisition tends to arouse controversy as to how the observations should be interpreted. The underlying problem is that we can't get inside the mind of, say, a chimpanzee to see what it is thinking. We have to interpret its actions.) Some of the most intriguing research on the use of protolanguage by apes is that of Sue Savage-Rumbaugh with her bonobo ape Kanzi. She has described the degree to which Kanzi has acquired protolanguage—using picture cards to communicate—in a number of books and articles, most recently the 1998 book *Apes, Language, and the Human Mind*, which she co-wrote with Stuart Shanker and Talbot Taylor.

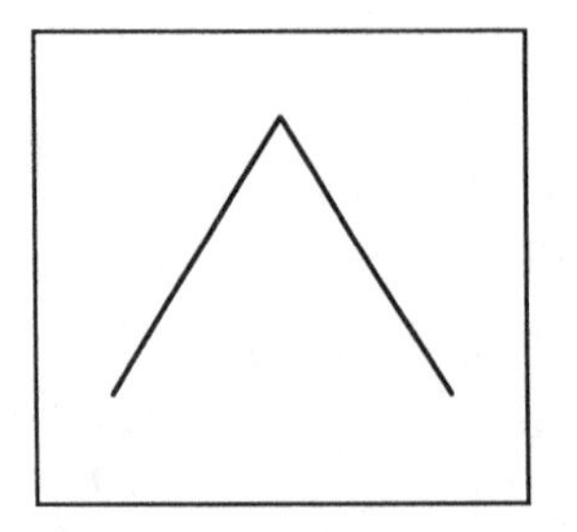

FIGURE 7.1 The Protolanguage Tree

The second stage in the evolution of language, the acquisition of syntax, occurred between 75,000 and 200,000 years ago. According to the standard theory, the need for ever richer communication led to the gradual enhancement of protolanguage with the various grammatical features that we find in today's languages. This theory may be correct. But in order to accept it, you first have to resolve two tantalizing puzzles.

First, if languages evolved gradually, they would surely have evolved at different rates. In that case, we should see contemporary languages of different grammatical complexity. But we don't. Among all the world's languages, not one has a simpler structure intermediate between protolanguage and full language. The phrase structure of the most primitive aboriginal tribe is virtually identical to that of the New York sophisticate.

Now for the second puzzle. The development of language took place when the only form of communication was face to face, when all the richness of body language—Shakespeare's "language in her eye, her cheek, her lip,... her foot...her wanton spirits, [and] every joint and motive of her body"—could be brought into play. But, as we noted earlier, practically all face-to-face, day-to-day communication about the here and now can be achieved using simple combinations of two or three content words, together with gestures. Thus, a sufficiently rich protolanguage is all you need. You don't need syntax—the grammatical structure of language—or the energy-expensive brain power it requires. So what additional survival benefits did the acquisition of syntax confer?

The benefits of full language come only when life has grown sufficiently complex to allow coordinated planning. For example, a well-developed protolanguage would enable one *Homo erectus* to say to another something like "dead mammoth" and point toward the north, or to say the *Homo erectus* equivalent of "fruit" while gesturing westward. And off they could trot in search of food. But protolanguage cannot convey a complex idea such as: "There is a dead mammoth there [gestures to the north], but yesterday there were tigers in the vicinity, so if we get there and find they are still around, we should be prepared to move on over there [gestures to the west] where there is plenty of great fruit with little danger." For that kind of communication, you need the syntactic structure of language. But such communication is only necessary when the creatures doing the communicating can generate such complex thoughts.

My view is that the standard account of the evolution of language is wrong. Language did not evolve *primarily* to facilitate greater communication. Rather, it arose, almost by chance, as a by-product of our ancestors acquiring the ability for an ever richer *understanding* of the world in which they found themselves—both the physical environment and their increasingly complex social world. The key development, I shall argue, is what I shall call "off-line thinking." (Neither the idea nor the name is original with me.)

Roughly speaking, off-line thinking is the capacity to reason in an abstract "what if?" fashion. By examining the nature of off-line thinking and the change in brain structure required to handle this new capacity, we shall see how it yields syntax (in the form of universal grammar) and the capacity for mathematical thought. Moreover, we shall see how the step from protolanguage to full language could have (perhaps must have) occurred rapidly, perhaps over a single generation, thereby leaving no traces of any intermediate form of language.

As I mentioned, the acquisition of language came at the end of three and a half million years of growth in protolanguage.[1] During this period, the human skull—and the brain inside it—grew to become nine times larger than is normal for a mammal of our body size, and three times larger than is normal for an ape. Since the adult human brain consumes 20 percent of the body's energy, even though it accounts for just 2 percent of body weight, its enlargement carried a heavy metabolic price tag. The growth in protolanguage (for the primary purpose of understanding the

1 The figure of three and a half million years comes from the archeological evidence of hominid skulls.

world, with communication secondary) provides one explanation for that expensive growth.

An account of the evolution of language, therefore, should begin with the growth of the hominid brain. But my main goal is to understand where our mathematical ability comes from. To do that, we need to start much further back in time. Let's start really early, with the demise of the dinosaurs. (In fact, identifiable precursors of mathematical thinking were arguably present even in the dinosaurs.)

HOW A LARGE METEOR LED TO SMARTER APES

Sixty-five million years ago, a large meteor, approximately six miles in diameter, smashed into what is today Mexico's Yucatan Peninsula, creating a massive crater some 186 miles across. The thick cloud of sulfurous dust thrown up into the atmosphere by the impact was carried around the earth by the winds, completely blocking out the sun's rays for several years. The planet's surface was thrust into darkness and intense cold. Unable to escape the cold or to survive for so long without food, the dinosaurs all perished. For 150 million years they had been the top of the food chain, and a dinosaur feared no creature apart from another dinosaur. They had maintained their dominant position and ensured their survival largely by virtue of size, strength, or speed. But when catastrophe struck, the stage was set for the emergence of a new strategy: acting smart.

Had you been present just before the meteor hit, it is unlikely that you would have guessed which of the many species living in the shadows of the dinosaurs would survive the long "meteor winter" and give rise to descendants who would eventually send one of their kind to the moon. In fact, you might not have seen this particular species at all. It was a small squirrel-like animal with a long, pointed nose, that scuttled among the trees and bushes of the dense tropical rain forests that then covered much of what is now Europe and North America. It lived on fruit and nuts, and its principal survival trick was to avoid being seen by anything that might eat it!

Hard as it is to imagine, one of these timid little creatures, clinging to a tiny niche of survival, in constant danger of being squashed flat by a passing dinosaur, would one day have a direct descendant who would walk on the moon. But evolution is fickle. The popular phrase "survival of the fittest" is a reasonably good description, provided you understand "fittest" to mean most suited to the conditions prevailing at the time. The moment that big meteor hit the planet, conditions changed dramatically. What for millions of years had been excellent survival profiles—namely, "big and fast" or "big and strong"—suddenly became death sentences. In the months and years following the meteor's impact, it was smart to be small, not dependent on lots of sunlight, capable of surviving on small amounts of dead vegetation, and able to hibernate with little expenditure of energy for months at a time. Neil Armstrong's great-grandmother by some six to ten million generations suddenly came into her own.[2]

When the dust finally settled back to earth and the sun's rays again penetrated the atmosphere, those small, squirrel-like creatures, having managed to survive the long period of cold and dark, found themselves in a new world—a world without dinosaurs. In the millennia that followed, they diversified into a great many new species, which came to dominate the woodlands and forests of the northern hemisphere. These species were the prosimians. Among their modern descendants are the lemurs of Madagascar and the galagos and lorises of Africa and Asia.

For 30 million years, conditions for life on earth remained fairly static, and the prosimian species prospered. Then, over a period of 2 million years, the earth's climate cooled. The surface temperature of the Pacific

2 The figure of six to ten million generations leading to Neil Armstrong assumes an average span of between six and ten years between generations. Incidentally, it's of interest to observe that if you were to take all the female members of that familial line stretching from a squirrel-like female at the time of the meteor all the way down to Neil Armstrong, you would have a population roughly equal to that of present-day New York City.

fell from 23°C to 17°C. That may not sound like much, but it was enough to cause major environmental changes. In particular, the extensive tropical habitat that covered much of the earth's land surface retreated to the band around the equator that we now call "the tropics."

Among the adaptations to this change in climate, one of the prosimian groups started to evolve into an entirely new line: the anthropoid primates (a group that includes the present-day monkeys and apes). With larger brains and more rounded faces, these early primates occupied what are now the equatorial regions of today's Africa, Asia, and South America. (At that time, what today are the separate American and African-Asian continents formed a single, connected whole.)

After the American and African-Asian continents separated, about 35 million years ago, the primates living in what is now South America evolved relatively little. Today's South American monkeys strongly resemble their early ancestors. The primates in Africa and Asia, however, continued to evolve. About 30 million years ago, the line split into two: the Old World monkeys—whose present-day descendants include the baboons, the macaques, and the langur monkeys—and the apes, from which we would descend.

For the following 20 million years, the apes were the dominant forest species, living off the rich fruits growing high in the trees where they lived. Then, about 10 million years ago, the climate changed again, becoming drier and around 10°C cooler. The rich forests gradually retreated, leaving the apes having to struggle for survival.

Around 7 million years ago, some apes started to venture beyond the forest edges and onto the open savannas. Exactly what triggered this move is not clear. (Indeed, as with most evolutionary accounts, there are some authorities who question whether it happened this way at all.)

One intriguing possibility is that the move was caused by the development of a single enzyme in the digestive system of the monkeys. According to this theory, the enzyme enabled the monkeys to eat unripe fruit, something the apes (and we, their descendants) cannot do to this day. This ensured a constant supply of ready food for the monkeys, but it reduced the food supply for the apes, much of which disappeared before it was edible for them. The only way the apes could survive was by descending from the trees and trying to make their way on the forest floor and at the edges of the forest. True or not, one thing I like about this theory is that it illustrates how evolutionary development often hinges on small changes.

In any event, the increased competition for food certainly led to a highly pressured existence for the apes. They may have acquired color vision at this time. Such a development—over many generations, of course—would have provided a clear survival benefit by making it easier to spot ripe fruits against the green background of the forest leaves.

Unfortunately, we have no direct evidence of exactly when the apes developed color vision. Certainly, at some stage, our ape ancestors became able to distinguish colors, and it seems likely that this occurred prior to our splitting away from the line leading to contemporary apes, which also have color vision. Whenever it occurred, it would have come at a price: color vision demands that the brain process far more complex visual information, requiring a larger organ with increased energy needs. Thus, whenever the change took place, it must have been at a time when the benefits outweighed these costs—after all, the apes did survive.

Even if it is true that color vision developed "in order to spot fruit better," we have to be cautious. Expressing it this way can give the false impression that "nature" solved a problem in the way that we solve problems each day. For example, at the individual level, if you have a toothache, you go to the dentist, who will fix it. With an evolutionary development such as color vision in the apes, however, it was not any individual ape that benefited. Rather, over many successive generations, those apes better able to distinguish colors would have a higher likelihood of surviving long enough to reproduce and pass on their genes. Thus, over time, as genes for color vision became more common, the ape *population* became more and more able to distinguish colors.

For us, the point of interest is not color vision itself, however, but the enlargement of the brain. A larger brain may have begun as a price paid for greater efficiency at finding edible fruit—a desperate and expensive gamble to cling onto the very edge of survival. But 10 million years later, descendants of those apes—humans—would make use of their large brains to do many things.

TO SURVIVE IN THE WIDE OPEN SPACES—ACT SMART!

Following their move to the open plains, the apes faced a new problem. Away from the shelter of the forest, they were much more prone to attack from predators—lions, tigers, leopards, hyenas, dogs, sometimes eagles, and occasionally other apes.

Three developments enabled them to survive. First, they grew bigger. Although very hungry predators will attack animals roughly their own size, mostly they prefer much smaller prey. An animal approaching the size of a modern human is often left alone. Second, they began to act smart—anticipating danger and taking steps to avoid it. Third, the apes started to form groups. By moving around in groups, the apes took advantage of safety in numbers—the fact that most predators prefer to attack a lone prey. For some of their descendants, group activity was greatly assisted by improved communication using the rudimentary form of language called protolanguage, about which I will say more in due course.

All three of these adaptations pushed the growth of the human brain; the latter two mark the beginnings of some precursors of mathematical thought. I shall examine these latter two survival techniques in turn—acting smart in this chapter and organizing into groups in Chapter 8.

First, then, acting smart. The first thing we should observe is that what we call "acting smart" depends on the agent and the environment. Let's consider a few examples.

The sunflower constantly turns its head to face the sun. Since it requires the sun's rays to live and grow, its behavior could be said to be "acting smart" from "the sunflower's perspective." The sunflower's turning is somewhat analogous to our heading for the refrigerator when we are hungry. But most of us would be very reluctant to talk about a sunflower even having a perspective, let alone describing its movements as "acting smart."

There are simple water-based bacteria that will move toward nutrients in the water and will swim away from regions that contain chemicals poisonous to them. From the point of view of the individual bacterium, that's surely pretty smart behavior. But as a fairly simple, low-level, chemical response, it is not what we generally mean by "acting smart."

Stomphia coccinea is a species of sea anemone whose ocean environment typically contains eleven types of starfish, only two of which prey on the anemone. If one of the nine non-threatening species of starfish happens to brush against a *Stomphia*, the anemone does not react. If one of the two predator starfish touches a *Stomphia*, however, the anemone recoils immediately. As with the bacteria, this is purely an automatic response to a particular chemical or collection of chemicals found in the two predator species but not in the other nine. But the *Stomphia*'s ability to distinguish between two categories of starfish—dangerous and not—is a cognitive ability that helps to keep it alive. Are we fully justified in maintaining that it is not "acting smart"?

Moving on to creatures with brains, when rooks see an animal coming too near their nest, they will pick up stones and drop them on the invader to drive the threat away. Is that "acting smart"?

A leaf-cutting ant that encounters an opening too narrow for the leaf it is carrying will maneuver the leaf until it can pass through the opening. Does that qualify as "acting smart"?

How about the octopus that figured out how to unscrew a mason jar in order to get to food inside? Acting smart or not?

Or consider the chimpanzee that, when presented with a banana suspended beyond its reach from the roof of the cage, thought for a while, then dragged a box from the other side of the cage until it was beneath the banana, climbed onto the box, and grabbed the banana. It is hard not to classify this as "acting smart" in the very same problem-solving sense that we would apply the term to humans.

The above examples might lead you to think there is a continuous spectrum of increasing degrees of "acting smart," with sunflowers at one end and humans at the other. I shall argue later that, whereas there is a spectrum, it is not continuous: there is a difference in kind between the chimpanzee and all the previous examples, and a second qualitative difference between chimpanzee intelligence and human intelligence. Moreover, those distinctions correspond to clearly delineated notions of "language."

The first examples of "smart behavior" I gave above are all simple stimulus-response mechanisms, albeit involving different levels of complexity. Some input from the physical world produces a physical response automatically and predictably.

The chimpanzee's solution to the banana problem, in contrast, requires a plan of action. The chimp not only *acts* smart, but *is* smart. Notice, however, that all the elements of the chimp's plan were physically present—to use a computing term, the chimp formulated its plan "on line," while still receiving visual input about the objects it was thinking about. In the taxonomy of abstract thinking that I introduced in Chapter 5, the chimp is capable of thought at level 2 abstraction. It's reasonable to assume that our ancestors were at about the same stage when they made the move from the forests to the open savannas. By looking for commonalties in the mental and communicative abilities of contemporary monkeys, apes, and chimpanzees (and humans), therefore, we can get some idea of those early savanna dwellers' mental capacities.

First, we have already seen that many creatures have a number sense and that this sense yields obvious survival benefits, ranging from recognizing which tree has the most fruit to knowing whether your group is outnumbered in a potential confrontation.

In addition to number sense, living in trees, with all that swinging from branch to branch, demands a good three-dimensional spatial sense, and survival on the open plains requires a two-dimensional spatial sense, including the ability to judge distance. I doubt that anyone would call such abilities geometric, but they are necessary prerequisites—the first steps, if you will—for geometric ability, the beginnings of a "geometry sense" analogous to number sense.

Another necessary component of mathematical thinking is an awareness of cause and effect. Since contemporary primates all appear to have such a capacity, it is reasonable to suppose that their (and our) ancestors in the forests and on the savanna likewise realized—if only in a restricted sense—that one thing can cause another.

Thus, as much as seven million years ago there were brains having some of the capacities necessary for mathematical thought. This is not to imagine that those early primates possessed anything like "mathematical ability." We have no reason to suppose that they had any capacity for reflective or abstract thought at all. Rather, my point is that mathematical thinking, as practiced today, makes use of mental capacities that were developed hundreds of thousands, and in some cases millions, of years ago. Doing mathematics does not require new mental abilities, but rather a novel use of some existing capacities. (Of course, strictly speaking, being able to use those existing capacities in a new way does in fact constitute a "new mental ability.")

For all these mental developments, however, arguably the most significant change in our early human ancestors at this time was not mental but physical: they started to change the way they moved. By coming to rely increasingly on their hind limbs alone for locomotion, they left their front limbs free for other purposes. By about three and a half million years ago, they were sufficiently different from the other apes that anthropologists place them in a distinct genus: the Australopithecines, the first of the hominids. With an appearance somewhere between that of an ape and a present-day human, these early hominids walked erect on two feet and used their hands as tools.

One particular consequence of the adoption of a fully upright posture that may well have proved crucial for our linguistic ability, and hence for our ability to do mathematics, is that we became able to utter consonants. It is our ability to separate a relatively small number of vowel sounds by means of consonants that makes it possible for us to produce the rich variety of sound-sequences by which we communicate. Without consonants, our oral communication would consist of little more than grunts. In short, we would not have language. (Remember that although I have suggested that representation drove the development of language, I did not say that communication played no role.)

Unlike all other primates, the human vocal tract curves downward at a right angle at the point where the oral cavity merges with the pharyngeal cavity. (In other primates, the vocal tract slopes gently downward.) This change in the human vocal tract was a direct consequence of the vertical positioning of the head over the spinal column, made necessary by our upright posture. If our head were tilted forward, like that of an ape or a chimpanzee, we would not be able to maintain our balance while walking on two limbs.

With the adoption of an upright posture, the hominid head became vertically balanced on the spine, and the tongue and the attachments of the larynx at the base of the tongue moved lower into the neck, creating a sharp angularity in the vocal tract and a smaller mouth. This change had the disadvantage that, unlike other mammals, we became unable to breathe and swallow at the same time. To prevent choking, our ancestors had to develop a sophisticated system to coordinate the two actions. By completely closing off the nasal cavity from the oral cavity—a procedure having the technical name of velar closure—we stop breathing for anywhere between half a second to four seconds in order to swallow. It is our ability to achieve velar closure that gives us our wide variety of consonants, and thus our language.[3]

3 Sue Savage-Rumbaugh suggests that the inability to form a broad range of recognizably different sounds may be the only reason other primates did not acquire language. If that is the case, then the communicative use of language could have played a larger role in its evolution than I am suggesting, although I still think that the major initial thrust was off-line thinking.

WALK TALL, THINK SMART, AND COMMUNICATE EFFECTIVELY

Following the Australopithecines, about 2 million years ago, came *Homo habilis.* Their appearance was still ape-like, although they were somewhat larger. Their brain was about 50 percent bigger than that of Australopithecine or of a contemporary ape (about 640 cubic centimeters compared with 440 cubic centimeters), although it was still just a third the size of a present-day human brain. But things were about to change in a big way.

Over the next million years, one line of hominids grew slowly even less ape-like and more like modern humans. About one million years ago, another new species emerged: *Homo erectus.* (It is not clear whether *Homo erectus* descended from *Homo habilis* or came from a separate, parallel line starting with the Australopithecines.) By far the most significant feature of the new species was its brain size, which was about 950 cubic centimeters, about twice that of a present-day ape.

Brain growth continued at a cracking pace. By the time *Homo sapiens*—the wise hominid—emerged around 300,000 years ago, it had a brain measuring some 1,350 cubic centimeters. That phenomenal growth had taken place in just 3,500,000 years.

As we have observed, given the enormous energy cost of supporting such a massive brain, the advantage conferred by its growth must have been significant. What drove this change? We simply do not know for sure. This is not to say that we cannot provide a plausible explanation. Indeed, this book is built around one theory that I personally find highly plausible. But unlike a lot of our knowledge of evolutionary development, where we have concrete evidence from the fossil record to support our theories, there is little to tell us what hominid brains were for. We have only the empty skulls that tell us how big those brains were. Once those brains were used to create tools and to paint and carve markings on the walls of caves, however, a record was left for today's anthropologists to examine. But the creation and use of tools was a very late use of brain power, and hence could not have driven that growth.

It is important to realize that evolutionary change does not have to have a clear explanation in terms that make sense to us. The popular nineteenth-century view of evolution as a process of "ascending a preordained ladder (with humans at the top)" was simply wrong. Individual evolutionary changes simply happen, caused by essentially random variations in the genetic code of a newly created member of the species. If some factor ensures that a particular genetic change results in members of the species being more likely to procreate and pass on their genes than members without the change, then the stage is set for that change to spread throughout the species. That crucial factor does not have to make immediate sense in terms of providing a rational survival advantage. Take the peacock's tail, for example.

At some stage in their evolution, some male peacocks started to grow larger and more colorful tails. Since large tails make it difficult for birds to move about and fly, they are clearly disadvantageous. Rationally, large tails should have been selected *against* and should quickly have died out. This is not what happened. A possible explanation is that female peahens viewed the extravagant plumage as a sign of strength—the bigger and more colorful the better—and hence preferred to mate with partners so endowed. If so, then large, colorful tails would be selected *for*, despite their obvious disadvantages.

Does a similar explanation work for the development of large brains in the hominid line? Did large brains develop for the simple reason that our ancestors preferred brainier partners? On reflection, such an explanation will not work, at least not in the same simple fashion it does for peacocks and peahens. Since brains are not visible, any preference could not have been for a large brain in itself, but rather for something that the owner of the brain did with that organ—presumably displaying clever behavior or being a better communicator. But now the explanation shifts to identifying the nature of that clever behavior, or to what advantage those big-brained individuals used their superior communicative ability. Thus, although this theory shifts the evolutionary selection factor from acting smart to choice of a sexual partner, we are still faced with our original question of what capacity or capacities large brains conferred on their owners that gave them a survival advantage.

My own view, on which this book is based, is that brain growth was driven by precisely the thing you would expect: larger brains enable us to act smarter and communicate better. Our task is to be more precise as to what "acting smarter" consisted of and to what end our ancestors put their improved communicative abilities. Moreover, the brain's size alone is not enough for smart behavior or better communication. We must also consider its structure.

THE DOG THAT CAN LEARN NEW TRICKS

We generally think of smart behavior, or intelligence, in terms of intellectual problem-solving ability, particularly problems that require highly focused, abstract thought (technically, "symbolic thought"—see the next chapter). But that kind of thought does not come easily, and relatively few people do it well. In fact, it is precisely the scarcity of such an ability that results in greater rewards for those who possess it—more attractive jobs, higher salaries, a better lifestyle, and the occasional Nobel Prize.

This ability also takes considerable effort to develop, requiring years of training (what we call education) and practice. In some instances, such as learning multiplication tables, much of that training involves trying to counter the natural tendencies—the natural "intelligence"—of the human brain.

In short, the kind of problem-solving ability that people generally react to with observations like "That individual is intelligent" (such as prowess in mathematics) is very much an aberration, even in the very society for which such an ability is arguably the most valuable. (Surely, our *Homo erectus* and early *Homo sapiens* ancestors had far less need for such "intelligence.") But evolutionary changes are manifested in the majority of any population, not the minorities. It seems unlikely in the extreme that the ability to solve intellectual problems was the driving force that selected for bigger brains.

If we want to identify that driving force, our best strategy would be to ask ourselves what mental abilities *all* human beings possess, not just those few whom we generally call "smart." The question we should be trying to answer is: what does *any* human brain do well? In what ways can every one of us be said to "act smart"?

We should guard against assuming that the ability in question requires either training or effort. In fact, *we should expect quite the contrary.* Evolution rarely, if ever, provides adaptations that require training or effort; rather, evolution leads to capacities that are well suited to a particular task. (To repeat, the adaptation did not come about in order to meet a particular need; rather, it was selected for because it did, by chance, meet some need.)

Thus, the mental ability we are looking for should be one that all human beings possess, and which requires no effort. Since no other species developed anything approaching the brain-to-body-size ratio of *Homo erectus*, let alone *Homo sapiens*, the key ability should be one possessed today only by humans. (If another species had that key ability, then that ability must not require such a large brain, so by definition it is not the ability we're looking for.)

Language is an obvious candidate. Except, that is, for one fact: human language is an extremely recent development. As far as we can tell, language—in the sense described in Chapter 6—arrived on the scene between 75,000 and at most 200,000 years ago. But hominid brain growth began around 3,500,000 years ago.

So our key ability is not language. On the other hand, it must surely be a precursor to language. The alternative is that the hominid line developed two independent abilities, one that drove brain growth for 3,500,000 years, the other that allowed language to spring onto the scene right at the end of that 3,500,000-year period. While it is technically possible for one line to be blessed with not one but two unique evolutionary developments that set the race apart from all others in so many ways, the odds against that happening are so remote that we should dismiss such a possibility from the start. Far more likely is that one development set the scene for the other: that language arose as a by-product of some more fundamental ability, the one that drove the initial brain growth over a 3,500,000-year period.

So what is that ability? If what I have been saying is correct, the answer should be staring us in the face, so obvious that we fail to see it, so commonplace and accepted that we do not pause to regard it as anything out of the ordinary. Something we simply take for granted. An ability we have had from the moment we were born. An ability that lies behind language.

Any ideas? Here's another clue. It was when I realized what this key ability must be that I knew everybody has the math gene—the ability to do mathematics.

If you are still puzzling over the matter, then almost certainly you are a victim of a view of the human mind propagated by the famous French philosopher René Descartes during the seventeenth century. The modern incarnation of Descartes' view is that the mind is a computing machine, which thinks by following a progression of discrete logical steps. According to this view, the key to understanding all of the other things people do with their minds—such as recognizing faces or understanding stories—is to express those mental activities in terms of logical rules.

As I argued in my book *Goodbye Descartes*, the assumption that all mental processes can be captured as logical rules is false. Moreover, it is precisely the falsity of that assumption that explains the failure of the many attempts to program digital computers to recognize scenes, handle natural language, and exhibit artificial intelligence. At the end of *Goodbye Descartes*, I proposed an alternative view of the human mind as a device for recognizing patterns—visual patterns, aural patterns, linguistic patterns, patterns of activities, patterns of behavior, logical patterns, and many others. Those patterns may be present in the world, or they may be imposed by the human mind as an integral part of its view of the world. (Some philosophers would argue that only patterns of the second kind are possible. My thesis is independent of that issue.)

Some of the patterns we recognize can be described using language (including specialized languages such as musical notation for describing certain aural patterns and mathematical notation for describing various mathematical patterns). Other patterns seem to defy all explicit description. For example, we generally have little difficulty recognizing a friend or relative we have not seen for several decades. In all probability, all details of the individual's face will have changed—indeed, our first reaction is to be surprised at how different they appear. And yet we recognize them nevertheless. Despite the many individual differences in their appearance, we know it is *the same face*. Recognition of a face seems to involve high-level patterns that survive any number of individual changes.

In *Goodbye Descartes*, I also argued that human expertise results not from learning to follow rules well but from acquiring the ability to recognize (and possibly create) a great many patterns and react accordingly. Rules, I argued, are useful for acquiring a new skill. Expertise comes about when the brain has adapted to that new skill—when it has learned to recognize the crucial types and make the appropriate response automatically and effortlessly. At that stage, rules are no longer needed.

Driving a car is an obvious example. Leaving aside criticisms of any one person's driving ability, the fact is that practically any human being can, and many do, become expert in driving a car. When we start learning to drive, we follow rules explicitly—and it shows! With a little practice, we become more fluent. We still follow rules, not explicitly but efficiently. Problems arise when we meet conditions we are not used to: typically, we revert to complete beginner status, having to think explicitly what to do. Eventually, after we have been driving for some time, we become so attuned to the various patterns and responses involved in driving that we can usually drive without giving the matter any attention, and our response when faced with an unexpected situation is instinctive and immediate. We no longer need the rules that guided the learning process.

It is perhaps not entirely superfluous to observe that human evolution can hardly have been guided by the development of the ability to drive a car. And yet we easily become highly skilled at driving. What evolution has provided us with is the ability to recognize new patterns and develop behavioral responses to those patterns.

According to the old saying, you can't teach an old dog new tricks. It would be more accurate to say that you can't teach an old dog very many new tricks. The same applies to young dogs, and to all other species as well, save one. We are the only dogs that can be taught more than a handful of new tricks. Indeed, we may learn a virtually unlimited number throughout our lives. This is an important aspect of the key ability that has led to our evolutionary success.

FOLLOWING THE LEARNING CURVE

The picture of human evolution that I am painting is of a two-stage development, beginning with the emergence of the hominid line three and a half million years ago.

In the first stage, which accounts for practically all of that time period, brain size steadily increased. It did so primarily in order to provide the owners of those larger brains with a richer view of the world (more patterns that could be recognized), a greater bag of tricks to aid survival (in the form of responses to particular patterns of stimuli), and a more effective means of communication. But the structure of the brain changed little. Development consisted primarily of *more* rather than *different*.

The second stage took place during the last 200,000 years, possibly as little as 75,000 years ago. Brains did not get bigger, but their structure changed. Whether or not we regard those structural changes themselves as anatomically major, their consequences in terms of brain activity were highly significant. They gave us symbolic (i.e., "off-line") thought (which I will explain in the next chapter), language, a sense of time, the ability to formulate and follow complex plans of action, and the ability to design and construct a dazzling and constantly growing array of artifacts. (These may appear to be very different abilities, but they all depend on the same fundamental capacity.)

To some extent, it seems fairly easy to understand how the first of these development stages—brain growth—took place. Yet if developing a richer collection of concepts about the world is sufficiently advantageous to counter the costs of a larger brain, why in the entire history of the earth has only one species followed that path?

Part of the answer could be that the large-brain, act-smart route to survival is a fairly desperate one that developed in a species that for most of its history lived on the very brink of extinction. All our hominid ancestors were easily outclassed by other species in size, strength, and speed. *Homo erectus* did not have the obvious advantage of lethally sharp claws or a large, powerful jaw with long, razor-sharp teeth. Nor did he have a thick skin, nor indeed any other anatomical defense against being ripped apart and eaten. He was master of no particular habitat and survived only by remaining highly adaptable, moving from place to place and from terrain to terrain in search of whatever food he could find—fruit, nuts, leaves, eggs, and occasionally meat from animals.

Success in this lifestyle depended on a rich view of the world—the ability to recognize a large collection of patterns. The greater and the richer our ancestors' understanding of the world, the greater their chances of survival—by foreseeing, and then staying away from, danger and by figuring out where the next meal would come from and acting accordingly.

An obvious instance where it would have been advantageous to recognize a larger collection of patterns would be in tracking. An injured rhinoceros might still fend off a *Homo erectus* band who happened to come across it. But if the band could track the animal over several days, taking advantage of clues such as footprints, broken branches, and animal droppings, then they could follow the rhino at a safe distance until it was sufficiently weakened that they could move in for the kill. The greater the number of patterns the hunters could differentiate, the more clues they could take advantage of, and the better they would be able to track their prey. The human ability to track prey, often over great distance for several days, is not shared by apes or chimpanzees, so it could well have been one of the factors that drove early hominid brain growth.

If our *Homo erectus* ancestors could also exchange information about the hunt using protolanguage utterances, they could coordinate their efforts, and so increase their chances of success.

In other words, our *Homo erectus* (and early *Homo sapiens*) ancestors were highly adaptable, nomadic scavengers who lived by their wits. Survival depended on their being smarter than any other species, and most probably their being able to communicate much finer information than any other species. They developed an intense curiosity about their environment and about all other species, a desire to *understand* and *explain* (at least to themselves) what they saw. This survival feature, unique to humans, is something we recognize even in young babies.

Given the energy costs of a large brain, the "acting smart" evolutionary path followed by our ancestors was almost certainly a very risky one, for which the most likely outcome was extinction. In the event, it worked, and those few thousand early hominids gave rise to some six billion human descendants, who have found out how to use their ancestors' desperate survival skills in order to establish a way of living unique among the creatures of the earth.

That, then, is my explanation for brain growth. The second stage of human brain development—symbolic reasoning—is much more problematic. We know from both the fossil record and linguistic evidence that it happened fast, and its consequences were dramatic. But why did it happen?

One possible explanation is that it was an accident. Of course, strictly speaking, all genetic changes are "accidental." But in general, a single, random change in the genetic code produces at most a tiny variation in the resulting life form, and it is only over time that many small changes accrue to make a major difference to a particular species. In the case of the acquisition of symbolic reasoning, however, the change was immediate and dramatic.

Personally, I don't have a problem with the "lucky accident" explanation. True, the odds against it must have been high, but we would not be here to reflect on the matter if it had not happened. Since we *are* here, the argument that such an occurrence is highly unlikely carries no weight. The situation is analogous to the familiar argument that the odds against winning a national lottery are so great that it's a waste of money to enter. This argument is perfectly rational (as a probabilistic argument), but it does not apply—after the fact—to the person who has just won the jackpot.

I shall consider a number of alternative explanations in some detail. Before I do, however, I should address a lingering issue concerning human brain growth. My explanation for the first stage of brain development does not involve problem solving and tool construction, both of which I claim came with the second stage of development. This runs counter to proposals that have been put forward by others. For example, P. V. Tobias, in his 1971 book *The Brain in Hominid Evolution,* suggests that the complexity of the brain developed along with its size, over a two-million-year period, and that this two-pronged development was accompanied by, and driven by, a corresponding increase in the complexity of the hominid way of life. On what evidence do I reject this and other proposals?

Simply this: the fossil record shows *no tools at all* from the *Homo habilis* period, during which time brains had already increased in size by 50 percent, and only a few very crude tools from the *Homo erectus* period, during which time the remainder of the brain growth took place. (Incidentally, the argument that perhaps our earlier ancestors used wooden tools that did not leave a fossil record has a simple counter. What about the tools they would have needed to manufacture the wooden tools? Even a wooden spear requires a stone tool for sharpening its end.)

Considering the dramatic explosion in tool use that followed the emergence of *Homo sapiens*, it is important to recognize just *how* crude were the tools of *Homo erectus*, particularly since some sources give a false impression. For instance, the best-known *Homo erectus* tool was the flint "ax," but the name suggests a more refined artifact than was the case. It was simply a sharp-edged flake that had been broken from a piece of flint. It was, we believe, held in the hand to remove the skin from dead animals. The survival advantage of such a device was significant. An ax wielder who came across, say, a dead rhinoceros would have been able to cut through its thick hide and butcher some of the flesh right away rather than having to wait until the carcass started to rot.

Like all the tools from the *Homo erectus* period, however, the flint ax could have been discovered accidentally rather than "invented" or "designed," and the idea simply passed on by demonstration and copying. After all, apes and chimpanzees will use rocks to break open the shells of nuts. Occasionally, a flint stone could have cracked, leaving a piece with a sharp edge that can penetrate skin. Of course, realizing the potential of the sharp edge was itself a significant act of intelligence, and we should not dismiss the ability to pass on various tool-related skills to the next generation. But when you consider that by the end of the *Homo erectus* period, the brain size of the majority of *Homo erectus* members was already in the lower half of the modern human range, while their collection of tools was still meager and crude, it seems clear that, whatever else that large brain was doing, its growth had not been driven by the design and use of tools.

It follows also that the initial brain growth was not driven by increased planning either, since the ability to formulate plans—as opposed to simple intentions to perform a particular action—would surely have resulted in better tools. Why? Because the mental processes required are essentially the same. The key ability for both planning and toolmaking is to be able to formulate a chain of "if this, then that" propositions.

For example, the individual who designed the first barbed spearhead must have gone through a thought process something like this. "Often, when I have speared an animal with a sharp stick, the stick falls out when the animal jerks about, and my dinner runs away. But if I cut a forward-pointing notch in the spear head, the spear won't be able to fall out, and as the animal moves about, the barb will make the hole bigger, so even if the animal does not die immediately and runs off, it will keep bleeding and die eventually, so all I need to do is follow it." It is hard to see how the first barbed spearhead could have been made without its design being reasoned out in such a fashion *in advance*.

This example comes from Bickerton's excellent 1995 book *Language and Human Behavior*. In fact, the argument I present here for a two-stage development of human mental ability is based on (and with one important exception essentially the same as) Bickerton's argument for a two-stage development of human language. Of course, all of the ideas that both of us are putting forward have been considered—and debated—by many others. Any novelty lies in the way we weave together those ideas to form a coherent story. Nevertheless, I owe Bickerton a large debt for at the very least helping to clarify my own thoughts on the development of human cognitive abilities.

At this point, it will be instructive to see how far *Homo erectus* had progressed along the path toward mathematical ability. In Chapter 1, I listed the mental attributes that contribute to this ability:

1. A number sense
2. Numerical ability
3. Algorithmic ability
4. The ability to handle abstraction
5. A sense of cause and effect
6. The ability to construct and follow a causal chain of facts or events
7. Logical reasoning ability
8. Relational reasoning ability
9. Spatial reasoning ability

Even back as far as *Homo habilis,* we have the beginnings of abilities 1 (number sense) and 9 (spatial reasoning). Likewise, given that all other contemporary higher primates demonstrate a sense of cause and effect (item 5), we can assume that *Homo habilis* had that ability too. The design of a barbed spear provides a clear indication that *Homo erectus* had ability 6 (constructing and following a causal chain of facts or events). The growth of societal living—a feature of *Homo erectus* (and later *Homo sapiens*) life that we'll examine in the next chapter—requires ability 8 (relational reasoning), because the mental mechanism that maintains such a lifestyle is the ability to remember and reason about relationships within the group.

The only abilities in our list that *Homo erectus* lacked were numbers 2, 3, 4, and 7. In fact, the key ability that is lacking is 4: the ability to handle abstraction—which I have elsewhere referred to as off-line thinking. Once you have that, all the other remaining items in our list follow almost automatically. For, as I shall argue in the following chapter, with off-line thinking you get language, and (as we saw in Chapter 3) when you have language on top of a number sense you get numerical ability, item 2. Moreover, algorithmic ability (item 3) and logical reasoning ability (item 7) are really just abstract versions of item 6.

Thus, with *Homo erectus,* our ancestors had many of the mental abilities that go to make up mathematical thought. Once a *Homo erectus* descendant came along who could form and handle abstraction, all the key mental pieces were in place for the subsequent development of mathematics.

Abstraction was the key step.

· 8 ·

OUT OF OUR MINDS

WHAT'S YOUR TYPE?

IN THE LAST CHAPTER, I said that, during the first stage of hominid cognitive development, our ancestors' spectacular brain growth was driven by the need for a richer view of the world, a greater repertoire of responses to particular patterns of stimuli, and a more effective means of communication. Here, I'll give some examples of what I meant.

First, however, we should agree on our terminology. Please bear with me. This book is not a technical research monograph, and as such it does not strive for scientific precision. In the present instant, however, a measure of precision will help to clarify an issue that lies at the center of the thesis I am putting forward.

The issue I want to address is: what exactly are meant by the words "pattern" and "type"?

As I have been using it here, the word "pattern" is not ideal. Not only does it suggest a visual pattern, but it can sometimes seem inappropriate to speak of a pattern being used to "classify objects in the world," which I want to do.

For example, most animals exhibit a pattern of gender: animals are either male or female. We can classify people in terms of gender (this person is male, that person is female), and we often modify our behavior toward each other based on this classification. The gender pattern is determined genetically and affects the roles that the individual animal plays in reproduction and the rearing of offspring. The male-female *pattern* is closely associated with the *properties* of gender, (being) male, and (being) female.

The rotation of the earth provides a pattern of night and day. Patterns in time are generally referred to as rhythms. The night-and-day rhythm is associated with the properties of night and day. The pattern influences the kinds of activities we engage in, and we sometimes use the properties (being) night and (being) day to explain our actions, by saying that "It's night-time" or "It's daytime."

Among the words used to refer to the things picked out by patterns are *property, category, type* ("What type of music do you prefer?"), and *kind* ("What kind of car do you drive?"). Other words denote the properties picked out by fairly specific kinds of pattern, such as *flavor, color, race,* and *size.*

Whatever the word used, the general idea is the same. We recognize patterns that split things up into two or more groups in a particular way. Moreover, it can be useful to us to regard those things as split up in that way. We may base our actions, or anticipate the actions of others, on which of the two or more groups a particular entity falls into.

Much technical work in the area of human cognition uses the word *type* to denote the concept referred to by any one of the terms *property, category, kind,* etc. Accordingly, that is the word I shall adopt in this book.

Thus, *male* and *female* are types (of people or animals), *night* and *day* are types (of environment), *sweet* and *sour* are types that apply to foods, *red, green,* and *blue* are types, *big* and *small* are types, *attractive* and *ugly* are types, *married* and *single* are types, and so forth.

Although this use of the word "type" seems strange in some cases, the idea itself is a very familiar one. It *should* be familiar by now, I believe, because the 3,500,000-year period of brain growth in the hominids was driven largely by the advantage of having an ever greater collection of types with which to classify things in the world.

Notice that the cognitive requirement for acquiring types is the ability to recognize similarities and differences: to realize that some things are similar—they are of the same *type*—and that other pairs of things are different—they are not of the same *type.*

Practically all living creatures have this ability to some extent: type recognition is the key to life. The bacterium in water that moves toward dissolved nutrients and away from poisonous water can be said to "recognize" the types *nutritious* and *poisonous* ("good" and "bad," if you like). The *Stomphia* anemone that we discussed in the previous chapter is able to distinguish between two kinds of starfish: *dangerous* and *non-threatening.* (In human scientific terms, the *Stomphia*'s single "dangerous" category contains two types of starfish, its single "non-threatening" category contains nine. Eleven types for humans, two for *Stomphia.* What constitutes a type is very much in the eye of the beholder.)

These two examples illustrate that it does not take a high intellect in order to "recognize" a type—at least in the sense that I just used the word. Indeed, it does not even require a brain. All that is required is that the creature modify its behavior in a systematic way according to the type.

At the same low level of activity, one could regard a simple thermostat as a device that can "recognize" the types *warm* and *cold.* After all, it reacts appropriately to the two types.

Regarding inanimate devices such as thermostats as "type-recognizers" does not affect my argument one way or the other, but it seems to me that there is a difference between animate and inanimate type-recognizers. The types that a living creature recognizes *matter* to it. Even for one as simple as a bacterium, the nutritious/poisonous distinction matters—indeed, it is a matter of life and death. But nothing matters to an inanimate type-recognizer. (Some experts prefer to speak of "differentiating a type" rather than "recognizing a type," on the grounds that "differentiate" does not suggest conscious cognitive activity. I tend to agree with this, and in my more technical writing I have used that terminology.)

In addition to the types *nutritious* and *poisonous,* which apply to ingestible substances, most animals recognize the types *friend* and *foe,* which apply to other animals. Adequate recognition of these types is again often a matter of life and death.

In modern society, medical doctors are people who have learned to recognize a great many types dealing with conditions of the human body: the type of having a cold, the type of having influenza, the type of being HIV positive, the type of having a high cholesterol level, the type of being overweight, blood type, and so on. Indeed, much of the training that doctors have to complete prior to being allowed to practice involves learning to recognize a great many such types, and learning to link each type of medical condition to a suitable *type* of treatment.

One evolutionary path that many creatures have followed is to increase the number of types that they recognize and respond to. Such species "progress" by successive generations responding better to various types than did their ancestors, and by differentiating *new* types (as the environment changes) that their ancestors did not differentiate.

Much of the time, such evolutionary "advances" comprise the acquisition (by a new generation) of new automatic stimulus-response links. These require no conscious effort or even cognitive activity of any kind. Humans, however, often do make a conscious effort to increase our body of types. This can be at the level of a particular individual, such as a trainee doctor, or it can be at the species level, as in medical research. Much medical research amounts to increasing the number of types of body conditions that can be recognized, and expanding the collection of types of treatment that can be effectively applied. This may involve refining existing types by splitting them into subtypes. Or we may discover that what were once thought to be separate types are really subtypes of the same type—a new common thread is observed. We refer to this enlargement of the collection of known types as "*progress* in medical research."

If you stop and think about it, it becomes clear that distinguishing types is the very essence of life, or at least of staying alive. Type recognition is so important that in many animals, large parts of one particular organ have evolved to handle types—to recognize types and to generate responses of the appropriate types. That organ is the brain.

Simple brains will recognize one or more types and in each case produce a bodily response of the appropriate type. Bigger brains have a greater repertoire of types, both types that they can recognize and response types that they can generate.

In the most simple brains, *all* the stimulus-type–response-type links appear to be "hard-wired." More complex brains, including human ones, also have *some* hard-wired responses. These hard-wired responses are all dedicated to ensure basic survival. For example, when we touch a hot surface with our hand, we retract the hand immediately, without thinking about it. But in addition to their hard-wired stimulus-response mechanisms, complex brains can also acquire new links through experience, i.e., through repeated exposure to a particular type of stimulus together with a way of evaluating the effectiveness of different types of response.

Animal trainers make use of such *adaptive* behavior to train animals. The trainer will issue a command. At first, the animal's response will be random. The trainer provides negative feedback when the response is not what she wants—either verbal or in the form of a minor physical blow—and positive feedback when the response happens to be correct—either verbal or by stroking, or perhaps by giving the animal a small item of food. Eventually, the animal learns to associate the command (the stimulus) with a particular action (the response) that brings a reward as opposed to a punishment. At that point, the reward and punishment need not be administered every time, but just used occasionally for reinforcement.

Sufficiently complex brains can also reverse the stimulus-response process, learning to perform an action of a particular type in order to produce an outcome of a desired type: "response" followed by "stimulus." For example, pigeons can be trained to depress a lever to obtain food. But notice that such learning is a consequence of the normal stimulus-response mechanism. The pigeon presses the lever by accident and food appears: stimulus followed by response. It hops up and down, but does not press the lever, and no food appears: again, stimulus followed by response. Eventually, its brain forms the link between pressing the lever and the appearance of food.

In all higher animals, including humans, the stimulus-response mechanisms that are hard-wired into the brain appear to be primarily lodged in a part of the brain called the amygdala, a small, almond-shaped region attached directly to the top of the spinal column. The amygdala is sometimes referred to as the "primitive brain," since it is present in all animal brains, and sometimes as the "reptilian brain," since it constitutes the entire brain of snakes and other reptiles. It is the amygdala that generates automatic responses to danger that help the animal to survive. It also produces often intense feelings of pleasure, fear, anger, and pain, and for that reason is sometimes called "the emotional brain."

Many "higher" animals, notably primates, have another area of brain surrounding the amygdala, the cortex. It is here that learned responses are developed and lodged. The largest part of the cortex, the neocortex, where conscious thought takes place, is the part that grew most during the 3,500,000-year period of hominid brain growth.

All parts of the brain are connected, and each can influence the other, although when danger threatens it is the amygdala that wins the day. For example, if a length of rope is thrown into a cage containing a monkey or a chimpanzee, the animal will instinctively flinch and recoil. The rope looks enough like a large snake to trigger the automatic survival response in the amygdala. Moments later, the neocortex recognizes that on this occasion there is no snake, just a harmless piece of rope, and the initial feeling of fear generated by the amygdala is quashed. People often exhibit a similar response.

Notice the order in which things happen in this example. The subject may well be engaged in conscious thought before the rope is thrown in, perhaps thinking about the next meal or trying to solve a mathematical problem. At that stage, the cortex is "in control." But once the eyes catch the motion of the rope, the amygdala immediately takes command, and the entire body goes into fear-driven survival mode. Only after the cortex has had time to assess the situation and declare that there is no danger does control revert back to the cortex.

Amygdala responses are automatic, instinctive, and much faster than the cortex. There are direct nervous system pathways to the amygdala from the eyes, ears, nose, and skin and out again to the body's motor controls. These "emergency channels" ensure that danger signals are acted upon without delay. This is why a thrown rope will produce an initial reaction, even though it might take only a fraction of a second for the cortex to classify the situation as non-threatening and cancel the response.

To summarize: in many species, an important function of brains—even the most simple brains—is to recognize certain types (e.g., "snake-like") and generate an appropriate response (i.e., one having the primary purpose of ensuring the creature's survival). The ability to recognize certain crucial types and to generate an appropriate response is hard-wired into the amygdala. The cortex, which can learn to recognize new types and generate suitable responses, is almost a second brain.

Like any summary, this picture is highly simplistic. Brains are highly complex organs that we do not yet fully understand, and *any* account that you read will oversimplify and gloss over details. But for our purposes, my account should serve well enough. In particular, it provides a plausible explanation of what capacities were acquired during the initial period of brain growth in hominids, and the advantage gained by that acquisition. Moreover, that explanation posits an incremental evolution in which, for most of the time, brains simply grew to do more of what they did originally.

The question remains: what caused the sudden apparently large jump in the capability of the human brain when it acquired language?

TELL ME ABOUT IT

So far in my discussion of brain development, I have said little about language or communication, save the unsupported observation that more effective communication was one of the factors driving brain growth. For species that live in societies, communication is the cement that holds them together. Thus, as our *Homo erectus* ancestors acquired a greater capacity for conceptualizing the world around them, they surely used that greater collection of types to communicate. This does not mean that they had language, with all the structure that entails. Indeed, they almost certainly did not have language until between 75,000 and 200,000 years ago. Prior to that, their communications were almost certainly in terms of what, following Bickerton and others, I have been calling *protolanguage*.

Protolanguage is the language equivalent of the type-object structure discussed above. In fact, Bickerton argues (and I agree) that once you have a concept of a *type* such as the type of a cat (i.e., an arbitrary cat), you already have a word "cat." By this, he does not mean a string of letters c-a-t that spells out the word, rather that such a type *is* a symbolic representation of a cat (i.e., of an arbitrary cat). Having such a symbolic representation is precisely what it means to say you have a *word* for cat.

Having a concept (or a symbolic representation) of a type is more than simply reacting to a type. We would not say that a thermostat that switches on the heat when the temperature falls below 21°C has a word meaning "cold," or that the animal that recoils when a rope is thrown down next to it has the word "snake" (or the word "rope," for that matter). Lots of creatures and devices respond in systematic ways to types—they "differentiate" types, to use the technical term. It is when a creature forms an abstract, symbolic representation of a type that Bickerton and I would say that it has a word for that type.

Having such a symbolic representation in your mind does not necessarily mean that you are able to articulate that word for someone else to understand. But it is not hard to see how such articulation could be developed, given a reliable means of producing a sufficiently broad range of sounds. If you take, say, an object in the world and a sound that is intended to refer to that object, you can (at least initially) accompany the utterance of that sound by a gesture that indicates the intended reference.

For example, if I repeatedly point at a cat when I make the sound that in written English we represent by the word "cat," it should eventually become pretty clear that this is what I intend my utterance to mean. Likewise, if I repeatedly smile broadly and look happy when making the sound that in written English we represent by the word "happy," my listener will probably come to understand that this sound means happy. Then, if I utter the two words "cat happy" together, while pointing at a particular cat, my listener stands a good chance of understanding me to mean that the cat I am pointing to is happy. Of course, things can go wrong at first, since the listener may assume more than one interpretation—maybe my utterance "happy" means smiling, which is not quite the same. But with constant repetition, such confusions will gradually be eliminated.

We know that this process can lead to effective communication, since we automatically resort to it when we find ourselves in a foreign country where we do not know the language and need to make our wishes known.

Given time, an initial exchange of the kind "Tarzan" (pointing to myself) and "Jane" (pointing to you) can lead to useful communication.

This observation gains significance when you recall that protolanguage has only content words, words that denote some object, action, or state in the world. It has no logical or grammatical words such as *the, a, to, by, from, because, and, if, then,* or *until.* Thus we can understand the long period of mental development that preceded the acquisition of language as one in which greater understanding (i.e., the acquisition of an increasingly rich collection of symbolically represented types) was accompanied by the introduction of sounds corresponding to those types. Understanding and communication developed together.

Once our ancestors had symbolic representations (i.e., they could recognize and refer to abstract types), it would be a small step to their producing simple protolanguage utterances of the form *object-type* or *type-object,* such as "Mother hungry" or "Mammoth come." (Word order is not important in such utterances: "Hungry mother" and "Mother hungry" mean the same, as do "Come mammoth" and "Mammoth come.") Utterances of this kind, where an object is ascribed to a type, are called *(simple) propositions* by linguists.

The pidgins that develop when people who speak different languages are brought together and must communicate to survive and work are also protolanguages. A much-cited example is the pidgin that developed in Hawaii in the late nineteenth century, when laborers from all over the world were brought in to work the sugar plantations. They communicated with simple object-property utterances using a mishmash of words from their native languages.

Although the vocabulary of the Hawaiian pidgin grew over time, it never developed into a full language with grammatical structure. But the Hawaiian-born children of the plantation workers spoke a fully grammatical language based on their parents' pidgin. Such amalgam languages are called creoles. The spontaneous emergence of a creole from a pidgin, in a single generation, is strong evidence that all human beings possess a built-in universal grammar.

Protolanguage propositions are also the kinds of utterances that infants utter during their first two years of life, before they progress (with remarkable speed) to full language. And insofar as anyone has met with any success in teaching apes, chimpanzees, and other animals to understand and use "language," it has also been restricted to the production of simple protolanguage propositions. Since these animals do not have the vocal apparatus to produce a sufficiently broad range of clearly differentiated sounds, researchers generally teach them to point to pictures or symbols printed on a card.

There are significant distinctions between children and apes in how they acquire protolanguage, how they use it, and what they use it to talk about. For one thing, the human infant acquires protolanguage with ease, whereas it takes years of training to teach an ape even a rudimentary collection of words. Also, a human infant will frequently initiate a conversation, and moreover the conversation may be about objects or events removed in space or time (what I have called level 2 abstraction). On the few occasions when a protolanguage-trained ape has been observed to start a conversation, it has been to express its own immediate wants or emotions or to indicate something in its immediate environment (level 1 abstraction at best).

The ability to use words to refer to objects and events not immediately to hand marks genuine *symbolic* reference, as opposed to mere "verbal pointing," and is a major step toward full language.

Bickerton makes the same point (*Language and Human Behavior*, pp. 51–52). (Where I have talked about types, symbols, and "a way of understanding the world," Bickerton talks respectively of categories, symbols, and lexicon.)

> The roots of the lexicon lie in two things: the capacity to sort objects into categories and the power to form associations between stimuli. Without

> categories, there can be nothing to attach symbols to, since linguistic symbols...do not relate directly to objects in the world, but rather to our concepts of the generalized classes to which raw objects belong.

He continues:

> Without associations between stimuli (rather than merely between stimulus and response), there would be no way in which symbols could be attached reliably to concepts.

By way of explanation, Bickerton reminds us of Pavlov's famous experiment with dogs. After a period of training where the ringing of a bell was followed by the arrival of food, the bell elicited the same response in the dogs—salivation in the expectation of food—as did the arrival of food itself. The dogs had learned to associate one stimulus, the ringing of the bell, with another, the arrival of food, and had transferred to the former stimulus their normal response to the latter. As Bickerton concludes, "It does not seem to stretch the meaning of 'symbolism' too far to say that for those dogs the sound of the bell served as a symbol for food."

I shall take up this further point in the next section, when I consider what exactly is a "symbol" in the linguistic sense.

SYMBOLS OF PROGRESS

On the surface, it seems simple enough: symbols are things that symbolize. The word *cat,* whether spoken or written, is a symbol that symbolizes a cat. The familiar circle-and-triangle figures on doorways in public places are universally recognized as symbols symbolizing women's and men's restrooms. And a particular cry from a vervet monkey is a symbol that tells the rest of the pack that a predator is in the vicinity.

Symbols *seem* simple, but appearances can be deceptive. As soon as you start to ask yourself what exactly constitutes a symbol, and how a symbol symbolizes, complexities arise. For one thing, practically anything can be a symbol. For example, we talk about particular people as "sex symbols" or fancy new cars as "status symbols." All it takes for something to be a symbol, it seems, is for us—or some creature—to use it to symbolize something. This at once shifts the question to: what does it mean for one thing to symbolize another? This puts the emphasis exactly where it should be, on the people and other animals that do the "symbolizing," rather than on the "symbols" themselves.

In fact, the question of what is involved in symbolizing is a difficult one and has attracted the attention of many philosophers, among them the American Charles Sanders Peirce. Now acknowledged as one of the greatest philosophers ever, Peirce was so far ahead of his time that his work was largely ignored during his life, and he was unable to secure a position at a university.

Peirce distinguished three levels whereby one thing can represent or depict something else. The simplest is what he called an *icon*. An object, action, event, or other entity X is called an icon for some other object, action, event, or other entity Y if there is a recognizable similarity between them whereby X brings Y to mind.

Many road signs are icons. For example, the (American) sign that warns motorists that there may be elderly people crossing the road shows silhouettes of two obviously elderly people walking. Similarly, the sign that indicates a cycle path has a silhouette of a person riding a bicycle. The advantage of using icons is that they transcend language. The meaning of a road sign should be "obvious," requiring no deliberate learning.

Peirce's next level of representation is called an *index*. An object, action, event, or other entity X is an index for some other object, action, event, or other entity Y if there is a causal connection between X and Y whereby X *indicates* Y. The essence of an indexical relationship is that the link is created by some physical or temporal connection between the indicator and the indicated. For example, the height of mercury in a thermometer is an index for the temperature, a weathervane is an index for the direction of the wind, and an automobile speedometer is an index of the car's speed. Human facial gestures and body language are largely indexical, representing emotional states. So are most animal alarm calls.

At the topmost level of Peirce's classification are *symbols*. An object, action, event, or entity X is called a symbol for some other object, action, event, or entity Y if there is an agreed convention whereby X represents (or *symbolizes*) Y, irrespective of the physical nature of X and Y. For example, a wedding ring is symbolic of the fact that the wearer is married. Nothing about a band of gold worn around the third finger of the left hand intrinsically represents a married state. The meaning is purely by convention.

Human language is almost entirely symbolic. For example, the relationship between the word *cat* (either spoken or written) and a particular kind of animal is a symbolic relationship. The symbolic relationships of language are grounded in culture and vary from one language to another.

Peirce's three categories of representation—icons, indices, and symbols—overlap, particularly the iconic and indexical categories. Representations form a spectrum, which he roughly but usefully divided into three categories. For example, when Pavlov's dogs learned to associate a ringing bell with the imminent delivery of food, the ringing bell was indexical of food and arguably verged on symbolic.

International toilet signs are largely symbolic, but they are subtly based on icons for male and female sexual organs. "No smoking" signs combine an icon—a drawing of a smoldering cigarette—with a symbolic element, the diagonal line that symbolizes prohibition. (Even the diagonal line can be seen as having an iconic origin, depicting a barrier that prevents access.)

Notice that, according to Peirce's definitions, "sex symbols" are not symbols but icons. Nor are "status symbols" such as fancy new cars or large houses symbols; they are indices.

Despite the fuzziness of the boundaries separating Peirce's three categories of representation, the distinction between indexical and symbolic representations seems by far the more significant. Indeed, it is arguable that only human beings have developed the capacity for true symbolic representation. Certainly we seem to be the only species that uses symbolic representation in an extensive and unlimited fashion.

A common counter-argument to the claim that symbols are uniquely human cites the behavior of the vervet monkey. The vervet, which lives in East Africa, has a vocabulary of at least three distinct shrieks and cries that warn of different dangers. It uses one call when it sees a python, another for a martial eagle, and a third for leopards. On hearing the python call, other vervets in the vicinity look around on the ground to locate the source of danger. The eagle call makes them descend from the trees and hide among the bushes. And they climb the nearest tree when they hear the leopard call.

The argument that the vervet monkey is capable of using its cry in a symbolic way goes roughly like this. Although these warning cries are essentially indexical, vervet monkeys have been observed to use their cries to deceive. For instance, a vervet monkey will occasionally make the leopard cry to send all the others into the trees, leaving it free to enjoy a particularly delectable item of food that it has discovered. By making the cry in the absence of an actual leopard, the argument goes, the vervet is using it in a purely symbolic way.

While I agree that such deception certainly indicates a degree of intelligence, I do not think it makes the use of the cry symbolic. On the contrary, the deception works precisely because all the other vervets, on hearing the signal, believe a leopard is nearby and dash for the nearest tree. Contrast this with our word "leopard". If I utter the word "leopard" in your presence, I may get your attention, but in the absence of any further information, nothing more will happen. You will certainly not assume that a leopard is roaming around loose. If we are in a zoo, you will most likely assume that I am indicating a nearby exhibit and glance around for it. Unlike the vervet, for whom the "leopard" call means one thing only—there is a leopard in the area—our word "leopard" refers to the abstract concept of a leopard and can be used in the complete absence of the real thing.

In other words, there is a big difference between the use of an indexical sign to mislead and a sign's being symbolic. Indeed, the vervet monkey's deception succeeds precisely *because* the representation is indexical.

To return to my main theme, Peirce's taxonomy of representations can help us understand how our *Homo erectus* and early *Homo sapiens* ancestors could have gradually acquired the ability to use symbols.

SYMBOLS FOR SYMBOLS

I have suggested already that the early hominids' gradual acquisition of more and more types gave them an increasingly rich view of the world, by allowing them to differentiate more and more regularities in the world. Yet modern humans can do more than that. Our world view includes a complex *structure* that connects types together.

For example, we recognize the type of all people. We also recognize various subtypes: male humans and female humans, children and adults, people with college degrees, farmers, the types of Americans, English, Germans, Mexicans, and so forth.

In addition to some types being subtypes (or refinements) of others, types form a complex, interrelated web. For instance, there are male adult Mexicans and female American farmers with college degrees. We make implicit use of this web whenever we communicate with one another; it is part of the way we understand the world.

The development of a type structure was a key step toward language. It may be where our ancestral line parted company with the other primates in terms of cognitive development. Certainly, no evidence indicates that any non-human animals that have been taught protolanguage have such a structure. They may have a few types or categories that apply to things in the world. But they are probably not aware of any connections between those categories.

Our *Homo erectus* ancestors likewise may have seen no structure connecting the different types that they recognized. We have no way of knowing. My own view is that in order for *Homo erectus*'s enormous brain to provide sufficient survival advantage to outweigh the cost of supporting it, the acquisition of a rich collection of types must have been accompanied by the acquisition of a type *structure*—i.e., their world view comprised not only types but connections between them. According to this picture, although they did not yet have full language, their protolanguage was already rich and powerful.

It is not clear whether those early hominids had also acquired the ability for symbolic representation and communication. They may have started out more like today's apes, but with a very large conceptual repertoire, able to recognize a wide variety of interrelated types and communicate about them in a largely iconic fashion. Over time, they could have developed indexical representations like the vervet monkeys' leopard cry.

Whether you regard this incremental process as culminating in true symbolic representation depends on how rigorously you want to define symbolic representation. My own inclination is toward a narrow definition that would exclude vervet monkeys and apes trained to use protolanguage. If a representation requires a direct, causal link to something in the immediate environment, I would call it indexical. True symbolic representation does not require such a link, and one symbol can denote or represent another. In the taxonomy of abstraction that I introduced in Chapter 5, indexical representation corresponds to level 2 abstraction, whereas symbolic representation corresponds to levels 3 and 4.

For instance, the physical symbol E is a visual symbol for a particular sound (an aural symbol), and both are symbols for the letter *e*, itself an abstract symbol that forms part of a *symbol system*. Combinations of symbols in that symbol system are used to form still other symbols: words.

Many creatures have the cognitive ability to recognize one thing as representative of another. Apes that have been taught to recognize and use signs even seem to distinguish between proper and common nouns. For instance, when they learn a particular sign for a banana, they apply it to any banana, whereas they realize that the signs that name various people who are familiar to them are unique to those individuals. But it takes months of highly repetitive training to teach an ape a mere handful of words, and even then they continue to make mistakes. Despite the "symbolic" appearance of the signs that they learn, it seems that they are at best forming indexical representations.

Contrast this behavior with that of a small child. As all parents know painfully well, a young child has an insatiable appetite for learning what things are called. Usually, the child needs to be told only once or twice in order to remember the name for the rest of his or her life. Then it's on to the next object to be named! Human beings, it seems, are born with (or acquire very early) an innate knowledge that things have names, and a built-in desire and ability to learn those names. This process of forming virtually instantaneous, often quite arbitrary, links, can in no way be described as forming indexical representations.

Thus, although we humans share with many other species the ability to regard one thing as representing another, only we can represent symbolically.

How did we acquire this unique ability?

CREATURES THAT CHANGE THE FACE OF THE EARTH

Some time between 75,000 and 200,000 years ago, *Homo sapiens* began to take off as a species. By 35,000 years ago, when the last Neanderthals had completely disappeared, our ancestors began to produce tools and artifacts at a prodigious rate that has never let up. Instead of adapting to the environment, they often adapted their immediate environment to meet their needs—not in the small and unchanging way of beavers who build dams or termites who construct mounds, but big time. They wore clothes, used fires to generate heat and light, and organized themselves into societies of even greater size and complexity. In so doing, they started down a path that has led to the insulated houses, roads, automobiles, airplanes, ships, snowplows, delivery trucks, room heaters, air conditioners, electric light, refrigeration, and all the other elements of the artificial mini-environments of our technological life.

No other species lives this way, or even comes close. Modern humans live in a way unlike any other creature on earth. Acknowledging this fact is not necessarily to put humans onto a pedestal. Our way of living may be terribly unstable. We have been around for only a tiny fraction of evolutionary time, and it is not at all clear that we will survive as long as our *Homo erectus* ancestors, let alone the 150 million years that the dinosaurs were around. Nor are we being species-centric when we acknowledge that our principal survival advantage—our big evolutionary trick—is qualitatively different from all other species.

Think of some of the things that only humans can do:

- We use language—not just protolanguage—to communicate about a great variety of objects, places, situations, and events in the world, often remote in time and space from where we are.
- We have full-blown symbolic representation.
- We can think "off-line," independently of outside stimuli and without giving rise to immediate action; we can think about events in the past, we can speculate about events in the future, and we can think about the passage of time.
- We use language to create fictional stories, to educate each other, and to entertain each other.
- We create a great variety of tools and both functional and purely symbolic artifacts.
- We create mini-environments to sustain human life where we would not otherwise be able to live.

- We formulate and follow often elaborate plans of future action. (Notice that, apart from hormonally triggered preparations for winter, animals show little evidence of advance planning.)
- We increase our understanding of the world and make decisions for action through logical reasoning.
- We are able to learn a wide variety of new skills during our lifetimes.
- We can count the number of objects in a collection.
- At least some of us can do mathematics.

In connection with the first item above, notice that human language is not just a particularly rich protolanguage. The words of the protolanguage that our pre-human ancestors used would have referred only to things having direct relevance to their immediate experience or having a direct influence on their survival. We, in contrast, can and do use our language to talk about *anything*.

This boundless scope is possible because of another unique feature of (genuine) language: it is compositional. We can combine a relatively few individual sounds to form a great many words, and we can combine those words to form a far greater number of phrases and sentences. In short, both in terms of what words may denote and the way it works, human language is qualitatively different from protolanguage.

There are two possibly unique features of human beings that I did not list above. The first is our ability to engage in conscious, self-reflective thought about ourselves. I left this ability off my list since there is no way of knowing whether it is unique to humans. Certainly, when we watch an ape, a chimpanzee, or a dolphin, we often get a sense of a self-aware being. Likewise when we gaze into the eyes of a dog, or even an octopus, we also have a sense of communication with another self-conscious creature. But these are only impressions. Since we cannot get inside the mind of another creature—or another human being for that matter—we can never be sure whether that creature is self-aware. (Observations of the way chimpanzees behave in front of mirrors do not demonstrate self-awareness, despite claims to the contrary.)

The second possibly unique ability that I left off my list is the capacity to form a "theory of mind." A theory of mind is when one individual imagines what it is like to be another and so tries to predict the other's behavior. Humans acquire this ability between four and five years of age. No other species seems to have this capacity to anything like the same extent as humans. There is some evidence that primates form a rudimentary theory of mind. The vervet monkey, for instance, produces its false leopard call knowing that it will have a particular effect on other vervets.

In another incident that is often cited, an adolescent male baboon was threatened by an approaching group of adults. Instead of running, it stood on its hind legs and stared into the distance, as if it had seen a predator. When the threatening adults all turned to see what the adolescent had noticed, the youngster made his escape (see Byrne and Whiten, "Tactical Deception of Familiar Individuals in Baboons").

Then there was the occasion when a female gorilla, moving with her group, noticed a partly hidden clump of edible vine. Pretending she had seen nothing, the gorilla stopped and started to groom herself. When the others had moved on, she uncovered the food and ate it undisturbed (see Whiten and Byrne, *Machiavellian Intelligence*).

Finally, when David Premack showed his chimpanzee Sarah a videotape of someone trying to solve a problem, and then presented Sarah with a range of tools, the chimp was able to pick the appropriate tool to solve the problem. To do this, Sarah must have first realized that the person on the tape was trying to solve a problem—in other words, the chimp had a sense of what it was like to be the person shown on the tape (see Premack and Woodruff, "Does the Chimpanzee Have a Theory of Mind?").

Leaving aside the two capacities just discussed, which other creatures besides ourselves may possess, we are still left with a substantial list of impressive abilities that are unique to humans. The fossil evidence for all of these abilities is less than 200,000 years old. Before then, there is nothing; after that time, there is an increasingly rich supply. If evolution is an incremental process whereby small changes that provide a tiny propagation advantage accumulate over long periods of time, how do we explain this sudden, dramatic change?

First, it is inconceivable that the many human abilities listed above are all independent. It's much more likely that they are all consequences of one key faculty. Thus our task is not to find separate mechanisms for each ability, but to find one mechanism that gave rise to them all.

There are two obvious possibilities. One is an incremental process in which small changes in the structure of the brain were accompanied by minor behavioral and cultural changes, and that those behavioral and cultural changes accumulated until a threshold level was reached, triggering major behavioral and cultural changes. The other possibility is that there was a sudden structural change within the brain. I'll give you two arguments that favor the second hypothesis.

First, if you accept that the fundamental language tree, outlined in Chapter 6, captures (at least in large part) the single combinational mechanism that generates the sentences in any language, then the sudden-change-in-brain-structure alternative follows on theoretical grounds. A brain either has that combinatorial mechanism or it doesn't. If it has, then you get full grammar; if not, you have only protolanguage. There is nothing in between. There is no simpler version of the fundamental language tree (Figure 6.5, page 160) apart from the property-object construction of protolanguage (the protolanguage tree, shown in Figure 7.1, page 170). If you remove any single component of the fundamental language tree, all you have left is the inverted V of the protolanguage tree.

This explanation is very much an analytic one, based on an assumption about how the brain might create sentences. For some solid observational evidence that language acquisition came through a sudden structural change in the brain, take the observable fact that all the world's languages have the same parts of speech and the same underlying universal grammar. If languages evolved piecemeal, we should see many different grammatical structures in today's languages, but we don't. There are protolanguages with varying degrees of vocabulary and there are full-blown grammatical languages, all having the same syntactic structure. In between, nothing. The most likely reason is that there can be no such intermediate language—syntax is all or nothing. In which case, it must have emerged suddenly. (This is not the same argument as the previous one. The first argument assumed that the human brain actually does generate sentences by something like the fundamental language tree. This second argument—which is due to Bickerton—is based purely on the observed grammatical structure of the world's languages, regardless of the mechanism used to produce sentences.)

The sudden-emergence theory does not prohibit the gradual development of *some aspects* of language. For example, the so-called attitude verbs—*believe, hope, suspect,* and so on—might have emerged later. Moreover, the initial *scope* of language might have been much the same as for protolanguage, namely, the physical world, and have embraced just those features of the world directly relevant to the survival of the *Homo sapiens* species. But that is a matter of scope, not structure. Likewise, the passive construction might have developed much later than the active. But in terms of the nested-clause structure of language (the fundamental language tree), the passive is just a variation of the active; the underlying structure is the same. Indeed if, like most linguists, we assume that the fundamental language tree (and its supporting theory) provides a universal grammar that captures the essential structure of language, then aside from these minor variations, there is no intermediate stage between protolanguage and full language.

But such quibbles aside, let me repeat the point that I made earlier: the acquisition of language and all the other uniquely human attributes that I listed was the result of a sudden structural change in the brain that gave us a new ability: the ability to think off-line.

This claim raises a number of questions:

1. What is off-line thinking, and how—in neurophysiological terms—did it come about?
2. How did off-line thinking produce all of the various human attributes listed on pages 213–214?
3. How did it result in the ability to engage in mathematical thinking?
4. How did it give us language?
5. How is it that all present-day human languages have the very same underlying structure, captured by the fundamental language tree?

Questions 3 and 4, of course, are included in question 2, but I have singled them out to emphasize their importance. I shall address question 5 first.

SEWING THE COMMON THREAD OF ALL LANGUAGES

There are a number of different theories of why all human languages have the same structure. I'll describe the three that appeal to me. Whichever one you prefer, it does not affect my overall argument about the origins of mathematical thought. Any of the three theories is consistent with the remainder of my account.

The first theory is what I call the "Linguistic Eve hypothesis." This says that the entire human species is descended from a single individual, born with a key structural change in its brain, or else we are all descended from the offspring of a single parent-pair, each of those offspring being born with that key structural change.

If this were the case, the common structure of all human languages would require no further explanation. It would simply be the linguistic structure that arose in that first linguistic brain. We would only need to explain how a structural change in a *single* brain (or perhaps the brains of the offspring of a single pair of parents) could produce the key source ability that gave an entire race the many abilities listed above.

A second possibility is that the structural change took place in many individuals at around the same time. If this were the case, how do we explain the fact that all human languages have essentially the same syntactic structure? One way would be to assume that the individuals in which the changes took place lived close together. Then we could argue that, during the long period of protolanguage growth, regular communication among those individuals fostered essentially the same changes in their brains. To complete the picture, we would then have to show that, although the step from protolanguage to language would prove culturally dramatic, in anatomical terms it was in fact a fairly minor change that was virtually forced by the communicative demands on the brain—what I call the "small change, big effect" phenomenon.

The third explanation is what I call the attractor theory.

THE ATTRACTOR THEORY

Once the human brain had reached sufficient size and complexity, physical and/or mathematical laws (presently unknown to us) meant that it *had* to change structurally in the way that it did. This would clearly explain the common structure of all human languages. Moreover, there would be no need to assume that the key change took place in one geographic region. Syntactic structure would be an inevitable consequence of the brain's increased complexity resulting from the growth in protolanguage.

This explanation puts syntactic structure into the same category as the animal coat patterns I discussed in Chapter 4, the elegant and sensuous "minimal surface" shapes adopted by soap films stretched across wire frames, the regular, hexagonal arrangement taken up by identical steel balls when optimally packed into a large container, and the spiral of the nautilus shell that results from its optimally economical way of enlarging its exterior armor as its body grows. Each of these regular, predictable, structured arrangements is a consequence of mathematical laws. In mathematical parlance, the final form is an *attractor* to which the system is drawn.

Unfortunately, not much can be said about this third possibility. We simply do not yet know enough about the mathematics of complex adaptive systems of even moderate complexity, let alone the massive complexity of the human brain. So let me first say what little I can about the attractor theory, and then take a closer look at the other two, "Linguistic Eve" and "small change, big effect."

The question is this: is the fundamental language tree an attractor, an inevitable, *mathematical* consequence of the three and a half million years of hominid brain growth? Personally, I see no reason why the answer might not be yes. Many developing or dynamic systems studied by mathematicians over the past century have turned out to have attractors. Why not the human brain?

Noam Chomsky, the father of modern linguistics, made this same observation a generation ago (*Reflections on Language*, p. 59):

> We know very little about what happens when 10^{10} neurons are crammed into something the size of a basketball, with further conditions imposed by the specific manner in which this system developed over time. It would be a serious error to suppose that all properties, or the interesting properties of the structures that evolved, can be "explained" in terms of natural selection.

Clearly, if syntactic structure is indeed "naturally emergent" once brains reach a certain size, density, and complexity, there is no need for us to postulate a sudden, possibly catastrophic change in brain structure. Syntax would be an inevitable, mathematical consequence of the growth of the hominid brain. This scenario would provide this book with a pleasingly tidy conclusion: the mathematical ability that comes from the human brain's acquisition of syntax gives us exactly the tools we need to understand how syntax emerged in the first place. Unfortunately, since we currently have no hard evidence to support such a theory, I must go on to consider other alternatives.

LINGUISTIC EVE

Surprising though the Linguistic Eve hypothesis might seem, there is some (controversial) biological evidence to suggest the existence of a single, female human ancestor some time between 150,000 and 200,000 years ago.[1] During the 1980s, researchers at the University of California at Berkeley and subsequently elsewhere mapped out the base-pair sequences of the mitochondrial DNA of large groups of women (over 100 in the first study, with larger groups in later studies).

Mitochondria, the constituents of cells that provide the cell with its energy, have their own DNA, independent of the cell nucleus. They are passed down from generation to generation, predominantly (it was once thought exclusively) through the female line, and thus provide the possibility of a fairly good "trace" of a direct female descendant line. By mapping and comparing the mitochondrial DNA of women from around the world and correlating the differences with the known average mutation rate for mitochondrial DNA, the researchers were able to construct a genealogical tree stretching back in time to a single source—a woman who lived between 150,000 and 200,000 years ago. Not surprisingly, this single female ancestor became known as "Mitochondrial Eve." Since the variation in mitochondrial DNA is greater in Africa than anywhere else in the world, that is where Mitochondrial Eve most probably lived, hence her alternative name "African Eve."

1 The existence of a common ancestor of all present-day humans is itself not controversial, and follows from elementary mathematical considerations of ancestral trees. At issue is how far back in time you have to go before the ancestral tree of the current world's population converges to a single node. The Linguistic Eve and the Mitochondrial Eve hypotheses say that we have a common *Homo sapiens* ancestor.

The time when Mitochondrial Eve probably lived, 150,000 to 200,000 years ago, coincides with the archaeological evidence of off-line, symbolic thought. So a Mitochondrial Eve is at least a candidate for a common symbolic/linguistic ancestor. But there is no reason to suppose that common ancestry in mitochondrial DNA has anything to do with the ability to think off-line or to use language, so I am not suggesting that language appeared on the scene along with Mitochondrial Eve. (Nor am I ruling it out.) The evidence for Mitochondrial Eve does, however, add credibility in two ways to the hypothesis that all present-day humans are descended from a single symbolic/linguistic ancestor.

First, the evidence for the existence of Mitochondrial Eve—controversial though it is—may counter skepticism about the suggestion that all living humans are descended from a single individual. Second, if our ancestral tree consists only of the descendants of a single individual as recently as 150,000 years ago, it becomes more likely that our early *Homo sapiens* ancestors after Eve developed brains that prepared the way for the eventual acquisition of language. (We shall see momentarily what such preparatory developments must have consisted of.)

Of course, because mitochondria are passed on predominantly through the female line, Mitochondrial Eve was by definition most probably a female. A symbolic/linguistic common ancestor could equally well have been male. Either way, since both the genetic and fossil evidence shows that *Homo sapiens* originated in Africa and migrated outward from there, it seems likely that any "Linguistic Eve"—or "Linguistic Adam"—likewise lived in Africa.

Incidentally, the existence of Mitochondrial Eve does not add an additional "improbable event" to the Linguistic Eve theory. There is no suggestion that this individual had *any* special new abilities. The most that we can conclude from the mitochondria evidence is that the genealogical tree of present-day human mitochondria seems to have a single root somewhere between 150,000 and 200,000 years ago. The only thing "special" about Mitochondrial Eve, if indeed she ever existed, is that she was a common *Homo sapiens* ancestor of all present-day humans.

The case of a common symbolic/linguistic ancestor, however—a Linguistic Eve—involves a definite mutation. Either that individual, or else some or all of her offspring, had a brain structure different from any brain the world had seen before. But what was the nature of that change?

It could have been a major mutation. Although most of us probably find this suggestion hard to accept—and certainly intellectually unsatisfying—we surely cannot rule out the possibility that our language, and all that follows from it, is a result of a widely improbable event. Like the individual who has just won the national lottery against billion-to-one odds, the fact that we are here, using language, makes the odds against language having originated with a chance event irrelevant.

Moreover, the history of life is full of chance events, with natural selection determining which of those events eventually takes root in the species. In the case of language acquisition, the growth of protolanguage and the structure of *Homo sapiens* society, both of which preceded language, would have provided an environment in which, once on the scene, language would be propagated by natural selection.

The alternative is that the mutation was not major; rather, the "mental circuits" required for syntax had already developed for some other purpose, and the appearance of Linguistic Eve was a result of a structurally minor change that was bound to happen sooner or later, particularly if the selective pressures on *Homo sapiens* brains were steadily pushing for increased representational and/or communicative capacity.

But this is just the "small change, big effect" scenario, the third of our three possible explanations for the origin of language. In other words, if a relatively minor change in brain structure resulted in the capacity for syntax, then *either* that change occurred in a single individual, who then passed it on to her or his children (or possibly it arose in the children of one individual), *or* the change occurred in many individuals, in which case you must explain why they all underwent essentially the same change at roughly the same time.

There are a number of specific "small change, big effect" scenarios for the acquisition of language. I'll describe two that I find attractive.

SMALL CHANGE, BIG EFFECT

The idea behind the "small change, big effect" explanation is that the key brain structure for handling syntax developed for some other purpose and was then co-opted for the new purpose of combining words to yield grammatical sentences. Takeovers of this kind are a common feature of evolutionary development. For example, the flotation bladders in certain species of fish became lungs in the mammals that they evolved into, and, in our own case, our hands were once feet. The well-known evolutionary biologist Stephen Jay Gould coined the verb *exapt* to denote the use of a capacity that evolved to meet one need in order to fulfill a different purpose. The idea is that *exaptation* contrasts with *adaptation*.

Gould illustrated the idea of exaptation with an example from architecture. Spandrels are the tapering triangular spaces with curved edges that you find in the upper corners of churches and cathedrals where a domed ceiling rests on arched supports that meet at a right angle. In many cases, such as the great dome of St. Mark's Cathedral in Venice, these spandrels are decorated with ornate artwork that fits harmoniously with the surrounding architecture. The effect is so striking that an observer would be forgiven for assuming that the spandrels were put there purely to carry the artwork. However, in order to support the ceiling, you need some form of supporting structure at the corner, and if that supporting element is designed to be curved in a way that blends with the wall arches and the dome of the ceiling, then you automatically get a spandrel. In other words, the spandrel arises from purely structural-architectural constraints. Its use to display artwork is an "afterthought"—an exaptation, Gould would say.

We can explain the emergence of bird flight as an example of exaptation. Some species of birds descended from a small meat-eating dinosaur, the Coelurosaur, which had large fins on its back for cooling—a definite evolutionary advantage in a hot climate. The fins evolved into the wings of the birds that descended from them. Let's follow a possible path from cooling fin to wing.

The larger the fins, and the more Coelurosaur can flap them, the greater the cooling effect and the better the chances of survival. Flapping becomes more efficient if the fins have an aerodynamic shape. If the fins are on the creature's side, rather than the top of its back, flapping the fins will create lift and take some of the weight off its limbs, allowing it to move faster. A creature that can move faster stands a greater chance either of catching prey or of escaping predators—or both. Thus, over many generations, we can expect fins to become more aerodynamic and to move steadily down the sides of the creatures.

Then, one day, this extraordinarily light-footed, side-finned dinosaur (do we still call it a Coelurosaur?) is running down prey or escaping a predator, flapping its fins, when it finds itself lifting off from the ground. Not far, of course. Perhaps it just misses one step. But that missed step gives it a greater chance of a good meal or of escape. From then on, succeeding generations will find they can "fly" greater and greater distances, until the day comes when a distant descendant of that first "step-skipper" embarks on genuine winged flight.

Can we find a similar explanation for the acquisition of syntax? Did the human language facility develop to meet some quite different purpose, unconnected with language?

The two situations are not quite the same. In the case of the acquisition of syntax, there seems to be a clear-cut distinction between having it and not; there is no intermediate stage between protolanguage and language. But with bird flight there is a continuous scale of possibilities between non-flight and flight. The point at which a long "hop" becomes a short "flight" is a matter of opinion. The same is true for the transition from fins to wings.

Thus, while the example of bird flight provides only an imperfect analogy for exaptation leading to language, it does suggest one possible explanation for the fact that all present-day human languages have the very same underlying structure.

If Coelurosaurs acquired wings over a long period of small changes to their fins, why do their descendants all have wings of the same shape? The answer is that the flat, blade-like shape of wings is strongly conditioned by the laws of aerodynamics. In order to create aerodynamic lift, a wing has to have such an underlying shape. Because wings evolved in many different species, we see a wide variety of wing shapes in today's bird populations. But because of the laws of aerodynamics, they are all variations on the same underlying shape.

Are there "laws" that similarly constrain all languages to have the same underlying structure? If so, what are those laws? This, of course, is just the attractor theory of language acquisition that we considered earlier.

Like the transition from fins to wings, the acquisition of language may also have benefited from anatomical proximity. The mental capacity to handle syntax—the fundamental language tree—may have initially developed for one purpose in a part of the brain close to the region that processed protolanguage. All that would then be required to exapt those brain circuits to provide syntactic structure to turn the protolanguage into language would be for a connection to be established between those adjacent brain regions.

We know from modern studies of the brain that it has a considerable degree of "plasticity." New neural connections can be made all the time, depending on the way the organ is used. This is how we learn new skills, from playing tennis or driving a car to speaking a foreign language. Since new neural connections form during the lifespan of a single individual, exaptation could surely lead to the transition from protolanguage to language in at most a few generations.

One intriguing possibility is that syntax is a consequence of the way we form complex sounds.

While some linguists—syntacticians—analyze the structure of sentences, others—phonologists—analyze the structure of words. Words, as we know, are built up from individual units called syllables. Just as syntacticians have found that all grammatical sentences of all human languages have the same underlying structure, represented by the fundamental language tree, the phonologists have discovered that all syllables of all human languages likewise share a common underlying structure, which can be represented by the syllabic structure tree shown in Figure 8.1. This tree shows how basic sounds—phonemes—fit together to form syllables.

The syllabic structure tree's role in phonology is different from that of the fundamental language tree in syntactic theory. For one thing, you may have to keep applying the fundamental language tree recursively in order to generate or analyze a sentence, whereas the syllabic structure tree applies directly to any syllable as a one-shot deal, i.e., it gives the vocal form (or "sound shape") of any syllable. Nevertheless, it is striking that the syllabic structure tree is exactly the same tree as the fundamental language tree. I shall say what conclusions I think we can draw from this observation in a moment. But first, let me elaborate a little on Figure 8.1.

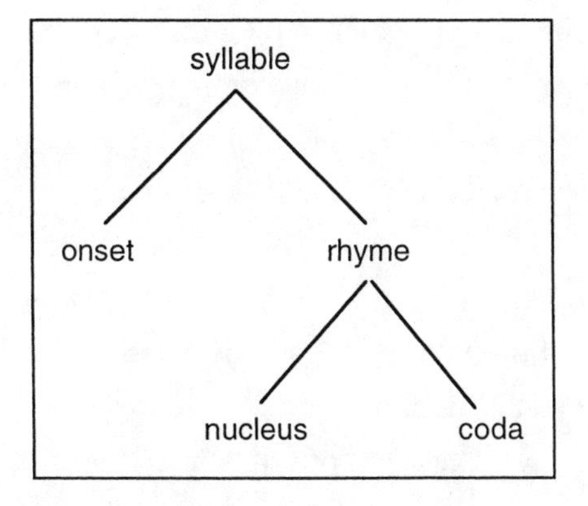

FIGURE 8.1 The Syllabic Structure Tree

The most basic vocal sounds are called phonemes. A modern human can make around three dozen phonemes, and it's reasonable to assume that our common ancestors had at most a similar vocal range. Thus, when *Homo erectus* first started to develop protolanguage some three million years ago (say), he had a vocal system that allowed him to articulate at most thirty-six different concepts.

In order to communicate about his growing collection of types (i.e., in order to vocalize his growing protolanguage), *Homo erectus* had to find a way to produce a much greater range of sounds. (We can assume that *Homo erectus* did in fact want to make meaningful utterances, since contemporary primates do.) One way to do so would be to put phonemes together to form more complicated sounds—what we now call syllables and words. (The most likely scenario is that syllables and words developed together, as meaningful utterances. The distinction between the two is largely a function of the way linguists *analyze* utterances. They find it convenient to view words as comprising a short string of syllables.)

If ever there was a strategy with a future, this was surely it. Out of our three dozen phonemes, the average modern adult creates a regular speaking vocabulary of around 15,000 words, each carrying meaning. We then use the same device—stringing things together—to create still more elaborate linguistic objects: sentences.

Of course, *Homo erectus* did not reason out this strategy. Rather, over many generations, the hominids gradually acquired a greater control of their vocal muscles—and forced a consequent change in their throat structure—enabling them to form an ever greater range of complex sounds. (We saw in Chapter 7 how *Homo erectus*'s adoption of an upright posture led to anatomical changes in the vocal tract that prepared the way for this development.)

As a result of these evolutionary changes, modern humans can successively widen and narrow the vocal tract with great rapidity and enormous precision in order to produce sounds that have noticeable variation in their perceived loudness (or sonority). Peaks of sonority—particularly the vowels—are the parts of an utterance we notice most, and they tend to dominate the patterns of sound that we hear.

Less sonorous sounds (consonants) naturally go immediately before or after certain sonority peaks, in a fashion that presumably reflects ease of neuromuscular control. Without these bracketing sounds, we could not produce anywhere near the range of recognizably distinct sounds that we do.

As our ears tell us whenever we travel abroad, different languages have different patterns of syllabic structure—they sound different. Nevertheless, they all share the underlying form captured by the syllabic structure tree. This shows an obligatory *nucleus,* surrounded by less sonorous margins called *onset* and *coda,* with the nucleus and coda forming what is called a *rhyme* alongside the onset. This is not a purely mental (i.e., symbolic) development, but a matter of bodily change and increased muscle control, just like learning to walk upright, increased manual dexterity, and greater accuracy in throwing. Presumably, the change took place over hundreds of thousands and perhaps one or two million years.

To return now to syntax, one possible explanation then for its origin is that it emerged when our ancestors' brains started to put syllables and words together in the same way they were used to putting phonemes together to form syllables.

How might this step have occurred? One possible mechanism was that the brain exapted the very circuits for generating syllables and used them to produce sentences. This would explain the suddenness with which syntax appeared. At a neurophysiological level, all that would be required would be to reach a threshold level whereby a direct pathway was established between the protolanguage area of the brain and the syllable-generating area.

Certainly the geography of the contemporary human brain is consistent with this hypothesis. Language is processed in the left frontal lobe—so the protolanguage of our *Homo erectus* and early *Homo sapiens* ancestors was presumably handled in the same region—and the production of vocal sounds is controlled by an area of the left frontal lobe called Broca's region. Moreover, damage to Broca's region not only leads to impairment of speech production but can also give rise to problems in handling grammar.

Thus, one way to view sentence production is as the culmination of a process that began by taking the mechanism that generates syllables (or syllables and words simultaneously) and, over many generations, using it to produce longer and longer sound sequences. (The region of the brain that was doing all the work would meanwhile have grown and undergone modifications to its structure.)

The hypothesis that syntax is an exaptation of syllabic structure has been advanced in various forms by a number of linguists, among them Peter MacNeilage, Michael Studdert-Kennedy, Björn Lindblom, Philip Lieberman, and Andrew Carstairs-McCarthy. In his most recent writing, Bickerton too suggests this as a possible mechanism.

Is this exaptation argument purely an artifact of representing syntactic structure and syllabic structure as trees? The answer is no. Let me explain why.

The fundamental language tree and the syllabic structure tree are theorists' mathematical devices to help them understand various aspects of language. But there is no suggestion that the brain contains such structures hard-wired *as trees.* Rather, the trees are a convenient way to represent on paper certain rules that are hard-wired in the brain *in some fashion.*

Most applications of mathematics to real-world phenomena are like this. For example, biomechanicists use differential equations to help them understand how people and animals walk, run, and throw objects. That does not mean that people and animals solve differential equations in their heads in order to carry out those activities. The differential equations are simply a mathematician's *way of describing* those various activities.

Mathematics, like any other conceptual framework, gives a particular *view* of any given phenomenon. Certain features are highlighted while others are obscured. But it is rare indeed for two things to "look the same" from a particular mathematical viewpoint and yet be unconnected.

Perhaps the best-known example is the appearance of the mathematical constant π in many areas of mathematics and physics. When π crops up in two seemingly unconnected places, it's almost never an accident; it generally indicates some deep, hidden connection between the two domains. As with π, so too with syntax and syllabic structure. I don't think the appearance of the same tree structure is just a fluke.

CROSSING THE SYMBOLIC RUBICON

Earlier in the chapter, I posed five questions:

1. What is off-line thinking and how—in neurophysiological terms—did it come about?
2. How did off-line thinking produce all of the various attributes listed on pages 213–214?
3. How did it result in the ability to engage in mathematical thinking?
4. How did it give us language?
5. How is it that all present-day human languages have the very same underlying structure, captured by the fundamental language tree?

So far, I have answered question 5—or rather I have given several plausible answers, any one of which is consistent with the remainder of my argument.

I now want to address question 1.

To try to understand what off-line thinking entails and how it could have come about, I shall review the evolution of brains in terms of a classification of "intelligent activity" introduced by the British psychologist Euan McPhail a few years ago. McPhail provided a three-point scale of intelligent behavior.

Practically all creatures exhibit McPhail's lowest level of intelligence: they can respond to certain stimuli. Call this stimulus-response (or S-R) behavior. You don't require a brain to produce S-R behavior, although the brains of the most primitive creatures have only S-R intelligence.

At the next level of intelligence, we find animals that can respond to a stimulus with another stimulus. Call this stimulus-stimulus (or S-S) behavior. All animals and many invertebrates exhibit S-S behavior. Learned S-S behavior involves the forming of indexical representations. Most probably brains developed initially as organs to facilitate S-S behavior. It is an effective survival strategy, since it allows an animal to adapt its response pattern based on past experience.

McPhail's third level of intelligence involves symbolic representation and language, and as far as we know, humans are the only creatures that have attained this level. Call it LI, for linguistic intelligence. (The reason for giving language such prominence in a classification of intelligent *behavior* will become clear presently.)

Like Peirce's three-point classification of representational modes into iconic, indexical, and symbolic, McPhail's classification is a rough division of a broad spectrum into three overlapping groupings. At least, the first two groups overlap, and we can understand how it is possible to progress up the spectrum by a long chain of incremental evolutionary developments. Like the step from indexical representations to symbolic, however, although some behaviors appear to lie on the borderline between S-R and S-S, going from S-S intelligence to LI seems to involve a discontinuity. So, once again we are faced with the same fundamental question: how do we get from one to the other?

Here is a purely fictional "explanation." Over three and a half million years, brains increase in size and complexity. As they do so, they develop an increasingly rich capacity to differentiate types of stimuli and forge links between those types. What began as an organ to generate physical responses to physical stimuli evolves into a device that can produce one internal stimulus from another. Over time, *Homo erectus* brains develop a wide repertoire of such stimulus-stimulus links. One might be willing to call this activity "thinking." But the subject matter of that "thinking" remains the physical environment. *Homo erectus*'s world—what he thinks about—is the physical world around him.

Imagine now (and this is the fictional part) that a second brain grows, parasitic on the first. This second brain has a similar structure to the first, except that its world is the first brain. Where the first brain receives its initial stimuli (its inputs) from the physical world, the second brain receives its stimuli from the first brain; and where the eventual responses (the outputs) of the first brain are physical actions in the world, the responses of the second brain are further stimuli to the first brain. We might then be inclined to refer to the activity of the second brain as "symbolic thought." Whereas the function of the first brain is to manipulate physical objects in the physical world, the function of the second brain is to manipulate symbolic objects that arise in the first brain.

To a first approximation, we can think of the prefrontal lobes as this second, "parasitic" brain. This part of the brain is a recent addition, and it's where much of language processing takes place. But given the degree of interconnections between all parts of the brain, this is far too simplistic. Strictly speaking, there is no parasitic second brain. That's what makes my account fictional. A more accurate way to think of the emergence of symbolic thought is that it arose when *the* brain—the first brain, if you like—*itself* developed the ability to function as the second brain. In other words, the brain became able to generate its own stimuli—to create and think about imaginary situations of its own creation, independent of any input from the physical world.

Let me explain these changes in physical terms. Although my description will undoubtedly be wrong in many details, in broad terms it is probably close to what actually happened.

Our present understanding of the way the brain works is that thinking (conscious or otherwise) involves fluctuating electrical (and electrochemical) activation patterns of networks of neurons. The simplest non-trivial example of a brain would have just two sets of neurons, one that accepts inputs from the outside world, another that generates output responses. Each input neuron is connected to several (maybe all) output neurons. The connections are of different capacities. The capacities of some connections may be variable, in that they become stronger the more frequently they are used.

Such a device, which is easily simulated on a digital computer in the form of what is called a neural net, produces stimulus-response behavior. Each pattern of inputs (different strengths of current going into each input neuron) generates a corresponding pattern of outputs. Where neural connections have variable capacities, the device can "learn" a new stimulus-response pattern if given the appropriate training. (This requires a feedback mechanism to adjust the capacities of the variable connections; see Figure 8.2.) For example, in the early 1980s, cognitive scientists David Rumelhart and James McClelland constructed a neural net with 460 input neurons connected to 460 output neurons that was able to learn how to form the past tense of a number of verbs, both regular and irregular (see Rumelhart and McClelland, *Parallel Distributed Processing*).

A more complex brain would have one or more layers of neurons (called "hidden layers") between the input and output neurons, mediating the flow between them. Such a brain, which is also easy to simulate as a neural net, is capable of more sophisticated behavior than a simple two-layer network.

Any mammalian brain is already far more complex than any neural net that can be simulated on a digital computer. The human brain, the mother of all brains, contains roughly 100 billion neurons, each directly connected to between 1,000 and 100,000 other neurons. The number of possible activation patterns is far larger than the number of atoms in the universe. The human brain has rightly been described as the most complicated known object in Creation.

Suppose for a moment that we were able to see inside the human brain, and that a neuron lit up whenever it fired, its brightness depending on the strength of the firing. Given an input pattern from the nervous system, we

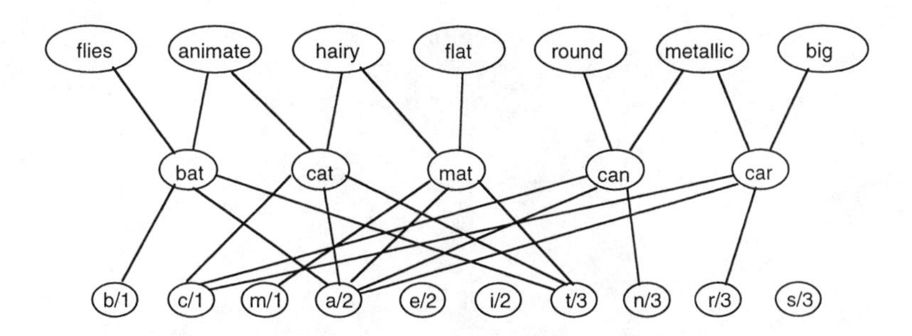

FIGURE 8.2 Part of a simple neural net that reads three-letter words and associates with them various properties of the objects they denote. The inputs (bottom layer) are individual letters in particular positions in the word. The middle layer represent units that the network recognizes as complete words. The outputs (top layer) are properties of the objects depicted by the words. The network is constructed so that three signals must reach a node in the middle layer in order for it to activate and send a signal to the output layer.

For example, if the word *cat* is input, by providing inputs to the nodes "c/1", "a/2", "t/3", then signals travel to reach one of the nodes on the middle layer. But only the node "cat" receives three inputs, so only it becomes active, sending signals to the nodes "animate" and "hairy". Thus, the network associates with the input of the word *cat* the output of the words *animate* and *hairy*.

would first observe a collection of illuminated neurons of different brightnesses spread all over the brain, with concentrations of brightly lit neurons in one or two areas (see Figure 8.3). This pattern would represent the input stimulus. We would then observe a veritable light show as current flowed (in parallel) from neuron to neuron. This is the brain processing (or thinking about) the input stimulus. If this activity resulted in a command to the body to perform a particular action, say to duck to avoid an approaching projectile, we would ultimately see a configuration of neurons light up that would cause signals to travel to the body's muscles, the effect of which would be that the body ducked down.

We would observe a similar sequence of electrical activity if we were likewise able to observe the internal workings of, say, any mammalian brain or of a pre-human hominid brain.

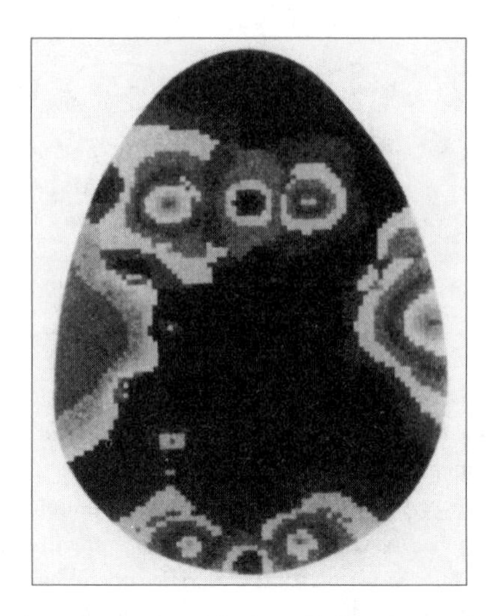

FIGURE 8.3 A top view of mental activity in a thinking human brain. Such a picture, called an electroencephalogram (EEG), is obtained by attaching electrodes to the subject's skull. It provides a map of the amount of electrical activity in each part of the brain. The lighter shaded regions are where the activity was most intense at the time the measurement was taken. In this example, there was considerable activity in the left side of the brain. The activity at the rear of the brain (bottom of picture) is the processing of visual signals received from the eyes.

I shall argue that the key difference between a human brain and that of any other creature, alive or extinct, is that the human can simulate any externally produced activation pattern to start the sequence, and run through the sequence without necessarily generating a bodily response. It is the activity of the brain originated by the brain itself, not caused by some external stimulus, and run without the automatic production of a bodily response, that I have been referring to as "off-line thinking."

Of course, not all brain activity requires an external stimulus. The brain of any creature is constantly active, monitoring, controlling, and initiating various body processes. Moreover, animals' brains initiate actions such as obtaining and eating food, and engaging in sex (although bodily signals to the brain are also involved in initiating those actions). Moreover, even when a human is engaged in the supremely off-line thought process of working on a mathematical problem, bodily responses are generated, perhaps the formation of a frown or the repeated tapping of a pencil on a desk. But these are not the kinds of stimuli or responses that I am referring to. Rather, I am thinking predominantly of the kind of brain activity that is generally occasioned by an external stimulus.[2]

Of course, an external stimulus can also cause off-line thinking. Indeed, it very often happens that some external event sets us onto a particular train of (off-line) thought. Thus, the description I have given is a simplistic one. The point I am trying to get across is this: an animal having S-S intelligence but not LI cannot initiate a particular pattern of brain activity that is normally occasioned by an external stimulus in the absence of such a stimulus. The appropriate initial activation pattern requires an input stimulus from the outside world.

In the case of a monkey, for example, the activation pattern that we might describe as "thinking about eating a juicy pear" can only be initiated by input from a particular juicy pear—say, sight or smell—or perhaps by a photograph of a pear. There is, unfortunately, no way of proving this, given present-day methods, since we have no way of knowing what a monkey is thinking about. Likewise, in the case of Pavlov's dogs, salivating at the sound of a bell, we have no way of knowing if they have any *thoughts* about an imminent meal. The most we can know for sure is that,

2 Dreaming is an interesting intermediate case, since dreams do appear to start spontaneously, in many non-humans as well as in humans. Despite considerable research, however, little is known about dreaming. It appears to involve random combination of parts of activation chains that have already been produced during waking activities. The repeat of parts of "meaningful" activation chains, which were once initiated by physical stimuli, may be what gives dreams their (recollected) content, even though there is a strong element of random firing involved. But since our only conscious knowledge of our dreams is the recollection that we have when we wake up, any meaningfulness may be illusory. And we have no way of knowing whether non-human dreaming, say by a cat or a dog, involves any mental activity meaningful to that animal. Given the problems involved in trying to understand what goes on in dreaming, it is far more likely that a greater understanding of thinking will cast light on the nature of dreams than that discoveries about dreams will help us understand thinking.

through training, one initiation stimulus (real food) can be replaced by another (a bell).

For an LI animal, however, the initial activation pattern that starts a particular activity sequence can be generated by the brain itself, without any external input. For example, as humans we can *imagine* a juicy pear, complete with sight, texture, smell, taste, and sound, and think about eating it, even when no actual pear is present. (Actually, the activity pattern activated by the sight of a real pear and that initiated by an imagined pear will not be exactly the same. Dieting would be easy indeed if imagined eating were subjectively indistinguishable from real eating.)

IN A WORLD OF OUR OWN

Off-line thinking is thinking about a world of internally generated symbols. Those symbols may correspond to real objects in the world, such as when we think about our distant relatives whom we have not seen for several years. Or we can think about things that have never existed, such as unicorns, or one of the totally impossible objects that featured in some of M. C. Escher's lithographs (see Figure 8.4).

Off-line thought may be initiated spontaneously by the brain or by direct input from a real object, or by some combination of the two, such as when climbing into our car begins a train of thought about whether to trade it in for a newer model. Our existing car is real—we are sitting in it; the new car that we think about acquiring in its place might not exist except as a mental amalgam of features we would like it to have.

As far as we know, no animals other than humans are able to think in this off-line fashion. Doing so is clearly a high-risk, high-gain strategy. One obvious gain comes from being able to reflect at length about past events and to plan future actions long in advance, thinking about various alternatives. This is an excellent survival strategy for a creature that is not particularly big, strong, or fast, has an easily penetrated outer skin, and

FIGURE 8.4 An impossible figure by the artist M. C. Escher, *Ascending and Descending* (1960). Our minds can imagine a figure that is physically impossible to achieve.

does not have sharp claws or a large jaw and strong teeth. An obvious risk is that a human engrossed in thought is a prime target for a passing saber-toothed tiger or some other mishap.

One way to reduce this risk is to ensure that part of the brain continues to monitor input stimuli from the world. This is exactly what happens. No matter what activity our "symbolic" neocortex is engaged in, the amygdala continues to do what it always did: keep its owner alive. If anything unexpected or threatening arises, it rouses the entire brain into taking evasive action.

For instance, we rely on this alarm system when we drive a car along a familiar route. Most of the time our mind wanders, perhaps reflecting on the day's activities or planning a camping trip. We are only vaguely conscious of our driving, or of what is going on around us. Then suddenly a small child runs out of a doorway toward the road ahead. Whatever we were thinking about is suddenly gone from our mind, and before we realize it we are braking fiercely to avoid an accident. All our attention is now on our driving.

With the amygdala constantly on the lookout for signs of danger, the potential drawbacks of off-line thought are significantly reduced, providing *Homo sapiens* with an opportunity to enjoy some of the benefits. Surely the most significant of those benefits was language. Off-line thinking automatically gives you full language. Or, more precisely, off-line thinking and language are two sides of the same coin. You can't have one without the other. This brilliant observation was made by Bickerton.[3]

Evidence that he's on the right track is provided by observation of those few unfortunate individuals who for one reason or another were not exposed to language during their formative years. Such people never become fluent in any language and have to make do with protolanguage throughout their lives. They also never acquire any significant ability to think off-line. A case in point is Joseph, an eleven-year-old boy described by neurologist Oliver Sacks in his book *Seeing Voices*. Born deaf, Joseph never heard language spoken, of course. In addition, he had not been

3 Bickerton's theory provided me with the missing piece of my own puzzle about the origins of the ability to do mathematics. Indeed, it filled the gap in my own thoughts with uncanny precision. In large part, this was because Bickerton's theory of language development tied in extremely well with my own thoughts about type recognition and type linkage (which I first investigated in depth in my 1991 book *Logic and Information*, itself a development of a theory of cognition begun by Jon Barwise and John Perry in the late 1970s and early 1980s; see Barwise and Perry, *Situations and Attitudes*). But the clincher for Bickerton's theory for me was his explanation of how it is that off-line, symbolic thought and language go hand in hand. His theory also helped me to understand how all the other uniquely human cognitive abilities that I listed earlier could arise from a single change in the structure of the brain.

exposed to fluent sign language. Thus, he had not been exposed to syntax. Here is what Sacks has to say about Joseph:

> Joseph saw, distinguished, categorized, used; he had no problems with perceptual categorization or generalization, but he could not, it seemed, go much beyond this, hold abstract ideas in mind, reflect, play, plan. He seemed completely literal—unable to juggle images or hypotheses or possibilities, unable to enter an imaginative or figurative realm. . . . He seemed, like an animal, or an infant, to be stuck in the present, to be confined to literal and immediate perception.

Joseph, in other words, could not think off-line, about things not in his immediate environment, be they real or imaginary things. And yet he did not have a genetic brain abnormality. The only unusual feature of his brain was that it was not exposed to grammatical language during his early childhood.

Besides showing that syntax is the main faculty of the brain that separates adult humans from infants and apes, the case of Joseph—and others like it—also shows that syntax is not entirely genetic. Human beings are born with a *potential* for syntax that makes it easy, natural, and instinctive for them to acquire syntax at an early age. But it takes exposure to language to trigger the development of syntax in each young brain. In that sense, we *learn* how to think abstractly and how to communicate using grammatical language.

What Bickerton did was provide a plausible explanation of why off-line thinking and language are necessary bedfellows. So what is Bickerton's theory, and how does it relate to mathematics?

THE SYMBOLIC MIND LIGHTS UP

As we have seen, Bickerton proposed that the 3,500,000-year period of hominid brain growth just prior to the emergence of *Homo sapiens* was driven by the development of protolanguage. This may seem to contradict my suggestion that the selective feature of brain growth was an increasing repertoire of types. But in fact the two approaches are virtually identical. The structure of Bickerton's protolanguage is essentially the type-object structure for understanding the world.

The type-object structure is about how we and other creatures understand and negotiate the world. A conceptual type structure allows you to better understand the world and find your way through it—the latter by responding appropriately to differing situations. Bickerton, on the other hand, was more interested in language. Protolanguage allows you to communicate type-object knowledge to others via single items of information of the "this object has that property" variety. (In my book *Logic and Information,* I called such simple items of information *infons.*)

For all the difference in emphasis, however, Bickerton's and my accounts essentially agree. I assumed that the type structure that provided the increased understanding and smarter behavior also facilitated better communication. Bickerton acknowledged that, in addition to supporting more effective communication, the growth of protolanguage provided a richer conceptual structure for understanding the world. Bickerton does suggest that it must have been the communicative function of protolanguage (or of type-object structure) that was selected for in the evolutionary mill, and here I am not sure I agree. I think better communication and an increased understanding of the world[4] were parallel driving forces. In any event, for three and a half million years, a growing brain provided an ever increasing repertoire of types (Devlin) or, equivalently, an ever larger protolanguage (Bickerton).

4 I am not talking about off-line, reflective thinking about the world here. By an increased understanding of the world, I simply mean the ability to recognize similarities between situations encountered, and to respond to a new situation in a fashion based on previous encounters with similar situations. This does not require off-line, symbolic thought and involves only iconic and indexical representation.

By at most 200,000 years ago, then, although they could not yet think off-line, our ancestors had a rich mental structure for what we might call "on-line thinking," i.e., intelligent action in response to increasingly subtle and complex physical stimuli, drawing on an increasingly large set of remembered experiences. What does this mean in terms of brain activity?

To answer this question we must first identify the brain activity that corresponds to (or produces) type recognition. As we have seen, the brain "develops" (or learns, if you prefer) by creating new pathways or by adjusting the capacities of existing pathways, so that different stimuli (or different collections of stimuli) cause different patterns of neuronal activation. Repeated exposure to the same kinds of stimuli causes various neural connections to adjust so that the brain can better recognize those kinds of stimuli—which means that the resulting activation pattern is stronger and more easily distinguished from other activation patterns.

In terms of electrical brain activity, then, types are certain well-developed activation patterns. (Equivalently, the constituents of Bickerton's protolanguage are certain well-developed brain activation patterns.) Moreover, when I speak of the acquisition of a larger repertoire of types or Bickerton refers to a growing protolanguage, we mean that the brain grows (i.e., acquires more neurons and more connections between them) and adjusts so that it can produce more activation patterns.

Now for our question of what kind of brain activity constitutes on-line thinking. A particular situation will set up a characteristic activation pattern. We generally refer to that activation pattern as the individual's cognition or awareness of the situation. Its precise form depends on the various types that the individual can recognize—the way it carves up the world. That initial activation pattern sets off a chain of further brain activation, which in non-humans (and often in humans) generally culminates in an instruction to the body to perform some motor action. This is the familiar stimulus-response cycle.

As far as we know, all animals except humans require a stimulus or group of stimuli from an actual, physical situation to start and maintain such a mental process. Indeed, the evolution of brains began with single cells or small groups of cells that could respond to a stimulus; all the intermediate complexity was added later. If the pre-human brain was purely a stimulus-response device, there would have been no reason for an activation chain to start spontaneously or to continue "off-line."

Let me stress that it is the physical environment that generates—through input stimuli—the activation patterns necessary to start and maintain on-line thought.

Off-line thinking, on the other hand, requires the brain itself to create the initializing activation pattern. To have survival value, this activation pattern must resemble one that would be produced by a real-world situation. Moreover, it must be possible for the resulting activation chain to culminate in a pattern that does not generate a motor action. (Off-line thinking generally leads to knowledge or an intention, say, not action.)

Before off-line thought was possible, Bickerton argues, the brain would have to have acquired the ability to produce a great variety of activation patterns—enough to combine several of them together to start and maintain a meaningful activation chain. In other words, protolanguage must have developed considerably before off-line thinking was possible. In my terminology, you need a lot of types before you can think off-line. In addition, you need a mechanism for combining them in a way that corresponds to the *structure* of real-world situations. What kinds of combinations are you likely to need?

Well, what kinds of connections between things do we encounter in real-world situations that give those situations their structure? Here are some: things stand in various kinds of relationship to other things, things act on other things, things combine to act on other things, things precede other things in time, actions cause other actions, actions prohibit other actions. But these are precisely the things that syntax gives you with its subjects, verbs, objects, and clausal structure. In other words, the combinatory machinery necessary to initiate and maintain off-line thinking is nothing other than syntax. When you get off-line thinking, you get full language, and vice versa.

That, in essence, is Bickerton's key observation. It is, as I have indicated, a brilliant insight.

Notice that his theory involves exaptation. The acquisition of off-line thinking—a faculty used to formulate plans and communicate complex ideas—was achieved by commandeering an ability that had evolved for a different purpose: on-line thought. On-line thought (equivalently, protolanguage) developed as a framework for ever more advanced stimulus-response behavior and (possibly) for communicating an increasing number of simple items of information. By finding a way to fake the conditions necessary to carry out such processes, the brain was able to bootstrap itself to a dramatically new level of ability.

And now we have answered all of the questions we set out to apart from one: how did *Homo sapiens* acquire the ability for mathematical thought?

As I have already indicated, I shall answer this question by showing how mathematical thought is simply a somewhat specialized form of off-line thinking. First, however, let's take a brief look at an interesting question that arises from Bickerton's account of the emergence of off-line thinking. Notice that his explanation does not require that the brain generate an exact replica of a real-world situation; just an activation pattern that is sufficiently like an externally generated pattern for the process to have survival value. The question is: which features of the world are absolutely necessary here—and hence will be incorporated into the syntax—and which ones can be "optional," present in some languages but not in others?

Surely, agent (subject), action or relationship (verb), and patient (object) are absolutely essential for any useful internal model of the world, and indeed these are embodied in universal grammar. A good case can also be made for at least some temporal (tense) distinctions. And in fact, all languages have grammatical rules that distinguish between past, present, and future. What other features of the world are incorporated into grammar?

The importance of reproduction to the survival of genes would explain why every language embodies gender in a fundamental way that involves inflections, demands agreement between different constituents of the sentence, and affects pronoun structure. (Many of these gender rules have disappeared from English, but persist in other languages.) On the other hand, it is also important to distinguish edible substances from poisonous ones and dangerous animals from benign ones, and yet neither of these important distinctions is reflected in the syntax of any language that I am aware of. Instead, they are made by having appropriate vocabulary.

Some features are incorporated in the syntax of some languages but not others. Turkish and Hopi, for example, have verbal inflections indicating whether a statement is based on personal experience or on information obtained second-hand. Many languages, English included, make no such distinction in their grammar.

In contrast, all languages incorporate into the syntax the distinction between one and more-than-one. Indeed, the singular-plural distinction is so fundamental that when linguists try to develop an adequate theory of grammar—the fundamental language tree—they find themselves having to put that distinction into the head of the sentence, the component of the sentence that governs all other components (see page 159). Think of the cognitive effort that must go into ensuring agreement between the plurality of nouns (generally reflected in their endings) and the inflection of the verb (usually reflected in endings and/or the principal vowel).

What makes the singular-plural distinction particularly intriguing is that it is arguably not the most fundamental one. The discussion of human and animal number sense in Chapter 2 suggests that the truly fundamental distinction is between one, two, and "many," where "many" means more than two. English makes this distinction using the lexicon rather than grammar. For example, we may speak of "one tiger" (1), "both tigers" or "a pair of tigers" (2), and "several tigers" (3 or more). But some languages make the three-way distinction grammatically, with verb inflections and agreement rules, pronoun structure, and so forth. Sanskrit, which dates back 3,000 years, is one of many such languages.

It is when the brain develops the ability to generate specific activation patterns that initiate and maintain meaningful chains of activation that (referring) words arise. What else would constitute the realization in the brain of the word "cat" but the brain activation pattern that stands for "a cat" in off-line thought? To put it another way, it is precisely when the brain acquires the ability to generate such a pattern spontaneously that we are justified in saying that the individual "knows" the word "cat." Thus, we have not only explained the origins of words but provided a neurophysiological description of what they are.

Now, at last, to the question: how did humans acquire our ability to do mathematics? Since my approach will be to show how mathematical thought is just a special kind of off-line thinking—which everyone can do—my answer will at once give rise to a second question: why do so many people think they cannot do mathematics?

·9·

WHERE DEMONS LURK AND MATHEMATICIANS WORK

THE STORY OF EVOLUTION is one of endless opportunism. Change occurs as a result of random "errors" (mutations) in copying DNA molecules. If a randomly introduced feature offers a certain survival or propagation advantage to the members of a species, that feature will tend to become steadily more prevalent until it is the norm. Sometimes, a feature that has evolved because it offered one survival advantage turns out to be useful in a quite different way. It then evolves further, driven by the new use. Eventually, it may no longer be possible to recognize the feature's original "function."

The ability of the human brain to produce and understand language arose in just this opportunistic fashion. On at least two occasions, nature took a feature that had developed for one thing and used it for another (the process evolutionists refer to as exaptation).

We have observed that brains originally functioned as stimulus-response devices: they recognized certain types and produced appropriate responses. Primitive brains—for example, those of reptiles or amphibians—have all their stimulus-response mechanism hard-wired into their structure. More complex brains, including human brains, have some stimulus-response mechanisms hard-wired, but they can also acquire additional stimulus-response connections through repetitive learning or training.

The human brain is by far the largest, compared to overall body size, of any animal that has ever lived. It is nine times larger than is normal for a mammal of our body size. It grew to that size over a period of about 3,500,000 years, ending around 300,000 years ago.

As we saw in Chapters 7 and 8, that growth was most likely driven primarily by the advantages of having a greater capacity to recognize types and being able to form increasingly rich associations between types of objects or situations encountered and appropriate types of responses. It is, however, likely that elements of this rich repertoire of recognized types were articulated as sounds, yielding an increasingly powerful communication system called protolanguage whose utterances were either single words or else simple object-property combinations. The early hominid line was a highly marginal one that in present-day terms would have easily been classified as an endangered species. *Homo erectus*'s brain was its main advantage, and any change in its brain that offered a better chance of survival would have quickly spread through the small population. Thus, once it was possible for hominids to communicate using protolanguage, they surely did so. Once they did, communication became another selection factor driving brain growth.

Indeed, once hominids started to communicate using protolanguage, communication may even have become the *dominant* selection factor driving further brain development. If so, that would give us another instance where a feature initially selected for one function (representation) turns out to provide a second, and eventually more significant, selection advantage (communication).

At the end of the long period of brain growth, somewhere between

75,000 and 200,000 years ago, the human brain acquired symbolic thinking. This was caused not by further brain enlargement but by a change in its structure. The brain acquired the ability to simulate the activation patterns normally caused by sensory stimuli, and to isolate from the body's motor centers the outcomes of the resulting thought processes. In short, the brain became able to think "off-line."

Off-line thinking could have developed only after the brain had acquired a sufficient number of types to support an internal "model" of the world that was adequate for reasoning. But a rich collection of types on its own would not be enough. The brain had to be able to represent, internally, a sufficiently rich world *structure*. The "skeletal structure" of the world that the brain simulated in order to think about the world off-line is what we now call syntactic structure, or syntax.

When you add syntax to protolanguage, you get language. Thus, in acquiring the ability to think off-line, the brain automatically acquired language. Since protolanguage was almost certainly being used for communication by that stage, language came into being as both a representational medium and a communicative medium.

This was another moment when nature—through natural selection—took off in a completely new direction. The advantages conferred by off-line thinking and language are so great that, within at most 200,000 years, *Homo sapiens* was living as no other species ever had.

Among the many things that off-line thinking permits human beings—and only human beings—to do are: we can communicate about a great variety of often remote objects, places, situations, and events in the world; we can think about past and future events; we can think about the passage of time; we can create fictional stories; we can create a great variety of tools and both functional and purely symbolic artifacts; we can formulate and follow often elaborate plans of future action; we can create mini-environments to sustain human life where it would not otherwise be able to survive; and we can increase our understanding of the world and make decisions for action by a process of logical reasoning.

In making these observations I am not being species-centric. Other species also have unique attributes—that is, after all, what evolution leads to. But no other species lives the way humans do. The lives of other creatures are centered on the basic necessities of staying alive: finding food and shelter, procreating, and (in some cases) raising their young. That's not at all the way we live.

We humans shape the environment to fit our needs, both on a large scale with our cities, our dams and reservoirs, our transportation systems, and our water, electricity, and sewage systems, and in the mini-environments we call our homes, complete with heating and air conditioning, electric light, water supply, sewage system, and so forth. No other creatures do this to anywhere near the degree we do.

As a result of the control that we exert on our environment, many of us devote only a small proportion of our efforts to staying alive. Instead, we do other things: we make and watch movies, write and read books, do science, learn new skills, paint, make and listen to music, build and go to amusement parks, entertain each other in a variety of ways, travel extensively for pleasure, and so forth. With the aid of books and the movie industry we have even turned the act of procreation into a form of entertainment, and with the aid of medical science into a casual pastime.

Much of this progress has been brought about with the aid of mathematics. And that brings us back to my original question: how did the human brain acquire the ability to do math?

The answer is: mathematics is an automatic consequence of off-line thinking. To put it in terms of our gene metaphor, the math gene and the language gene are one and the same. That means that *Homo sapiens* acquired the math gene between 75,000 and 200,000 years ago, when our species acquired language.

Once acquired, however, the ability to do mathematics lay largely dormant for over 75,000 years, until the Sumerian, Babylonian, Egyptian, and Chinese societies started out along the mathematical path that the human race has been treading for the past 5,000 years. Why did it take over

75,000 years before humans started to make use of their latent capacity for mathematical thought?

One perfectly reasonable answer is that it just did, that's all. Another is that society had to reach a certain level of organization before mathematical ideas and the mathematical way of thinking could be beneficial. The trouble with this answer is that it implicitly assumes an evolution of society toward one that includes mathematical thinking. Anyone reading this book lives in a society where mathematical thinking is one of the main driving forces behind many aspects of life. But I see no reason why things had to turn out that way; they didn't for most of the history of *Homo sapiens.* Moreover, the development of mathematics over the past 5,000 years has been carried out by a very small percentage of people, and had those particular individuals not lived, we might still not have mathematics.

The same cannot be said of language. Human beings found a particularly important use for language the moment it arrived, and they have been using it for that purpose ever since. What is that purpose?

Answering this question will, in fact, lead us to the explanation of why the language gene and the math gene are one and the same.

HE SAID, SHE SAID

A human baby utters its first word about eighteen months after it is born, often "daddy," because most mothers use the word frequently in talking to their babies. By the time it is two, the baby has a vocabulary of about fifty words. By age three that total has jumped to a thousand or so words that it knows and can use, and it has started to string two or three words together to form simple—but generally grammatically correct—sentences to make requests. Between the ages of three and six, the child's rate of learning is the highest it will ever be for any kind of knowledge acquisition: it learns an average of eleven new words a day, so that its (comprehension) vocabulary at age six is around 13,000 words, already approximately 20 percent of the 60,000-word (comprehension) vocabulary it will have at age eighteen.

The child will use language for a great many things over its lifetime. As we have already observed, language is a powerful and remarkably versatile tool. We can use spoken language to convey factual information, give commands, ask questions, create social and legal entities (marriages, companies, contracts, states of war, etc.), convey our inner emotions, comfort, impress, intimidate, amuse, entertain, express anger, teach, play games, thank, court a partner, seduce, manage companies, buy and sell things, help coordinate our actions when we are engaged in a joint project (such as land on the moon), and a host of other things as well. With written language, we can do most of the above, as well as store information and create lasting records of various events.

I'm sure you could extend this list. Singing? Yes, of course. Poetry? That too. Doubtless you can come up with further examples. Yet the chances are you will not mention the one thing that spoken language is used for most of all. Nor will you find this use of language mentioned in most books on linguistics—certainly not those that concentrate on the theory or reflect on language from a philosophical perspective.

The reason why this particular use of language so often gets overlooked is that it is so obvious and everyday. The one thing that most people use language for most of the time is gossip.

What was that you said? That doesn't apply to you? You are so busy using language for other things you don't have much time for gossip? You may think that, but a lot of evidence says otherwise. For many years now, two groups of linguists, called sociolinguists and psycholinguists, have been going around with notebooks and tape recorders, discreetly taking note of what people talk about and how much time they spend doing so. They have found that, on average, roughly *two-thirds* of all conversations are taken up with social matters—who is doing what with whom and whether it's a good thing or not, problems within relationships and how to handle them, and problems and activities at work, school, or in the family. In short, gossip.

Of course, for meetings whose intended purpose is not social, such as committee meetings of various kinds, the figure can be lower than two-thirds—but in general it is not *that* much lower. And at social gatherings in bars, cafés, restaurants, or chance meetings in the street, the talk is of little else. Even when barroom talk turns to politics or sport, much of it is essentially gossip about the people involved. And if you think most of the talk at a scientific conference concentrates on science rather than gossip about the people who do it, either you haven't been to a scientific conference or else you have simply done what most linguists do when studying language: completely overlook its primary use. What people talk about is mostly other people.

The same proportion crops up when you measure the shelf space taken up by fiction in a typical bookstore. The average is around two-thirds. What is fiction other than packaged, imaginary gossip? And the most popular fiction, by a large margin, is romantic fiction—imaginary gossip about relationships between men and women.

What about newspapers? Again, an average of two-thirds of the column inches in your daily paper is devoted to "human interest" stories (a journalistic euphemism for gossip), with a range from 75 or 80 percent in the "locals" and "populars" (such as *USA Today*), down to a still substantial 40 percent for the "heavies" such as the *New York Times, Wall Street Journal*, and *Washington Post*.

The figures for television news are similar. The biggest "news" stories on American television news in recent years were the O. J. Simpson trial and the Bill Clinton–Monica Lewinsky sex scandal.

What do we find when we put our feet up and turn on the television not for information but entertainment? Mostly gossip. Yes, there are sports broadcasts, music concerts, political discussions, and documentaries. But most of what we watch is about the lives of other people, either real or fictional. Some of it is well-made movies and quality drama shows that tell a good story. But the programs that attract the greatest viewing audiences are the soap operas (fictional gossip about supposedly ordinary people like ourselves), chat shows (supposedly real gossip about the everyday lives of the rich and famous), and the "ordinary people tell all" programs such as *The Jerry Springer Show.*

What this prevalence of gossip tells us about ourselves is that almost all of us are interested in the lives of our fellow humans. Not all six billion fellow humans, of course. But as soon as we know enough about someone, our interest is sufficiently aroused that we want to know more. The only "defense" against this human compulsion is to avoid becoming interested in the first place—as can be attested by any "too busy for gossip," "live life to the fullest," jet-setting business executive (BE) who has been laid up in bed sick for a few weeks with only daytime television to while away the hours. At first, the bedridden BE will resist the soaps, flicking the remote control to find something else. But eventually, the BE will find himself watching a scene or two of *Days of Our Lives, General Hospital,* or whatever. Within days, the hook is firmly embedded, and BE looks forward eagerly to each new episode. He knows that the dialogue is poor, the acting wooden, and the plots melodramatic. But that doesn't diminish his desire to find out what happens next to those characters.

This natural tendency to want to know more about the lives of people we know—even if they are fictional—was brought home to me dramatically about twenty-five years ago when an American television company introduced a series called *Mary Hartman, Mary Hartman.* Developed as a satire of the daytime soaps, it was broadcast late at night for the educated sophisticates who watched *Masterpiece Theater.* My wife and I were living in Canada at the time, and we found it hilarious—a delicious send-up of the soaps we never watched. About a month into the series, we realized we were hooked—as were millions of other viewers all over North America. The soap satire had become a soap!

The same thing happened with a subsequent satirical series called simply *Soap,* which went a step further than *Mary Hartman* by making the plot line even more ridiculous.

From an evolutionary viewpoint, what is the purpose of gossip?

The common view is that it serves no purpose whatsoever, that it is a distraction from the "more important" things in our lives. But from an evolutionary standpoint, that can't be right. Natural selection is a hard taskmaster. For those of us who live in today's highly civilized societies, gossip may seem to be merely a distraction, but the same cannot be claimed for Third World peoples, for whom life is a desperate struggle for survival. And yet studies have shown that they gossip just as much as any Western sophisticate. (Actually, in terms of face-to-face gossip, they do far more, but when you factor in our packaged gossip—novels, movies, and television—then things even up.) If people the world over, both now and in the past, use language primarily to gossip about each other, there must be a very good reason. What is it?

Let's go back to when *Homo sapiens* first acquired language, 75,000 to 200,000 years ago. What were their lives like? How did they ensure their survival?

Our ancestors, like ourselves, were highly social creatures. They formed groups, seeking safety in numbers. Of course, many other species form groups, and none of them have even protolanguage, so groups living alone cannot be the reason for gossip. Nevertheless, because of the survival value of planning to our early *Homo sapiens* ancestors, they doubtless used language (not just protolanguage) to organize themselves in stable groups and cooperate in performing increasingly complex tasks. But that's not gossip. I suspect the road to gossip began as follows.

With off-line thinking, it became possible to think about things not in the immediate environment. Consequently, group members could carry out group activities—such as following a plan worked out in advance—even when the group was not all in the same location. The rapid transformation of *Homo sapiens* from endangered species to a population of six billion a mere 75,000 to 200,000 years later (a blink of an eye in evolutionary terms) can be attributed as much to such increasingly sophisticated group activity as to any other factor.

But what would motivate an individual to act on behalf of the group when the group was far away? Something surely must have. If each individual, when away from the group, always put individual comfort and safety above the group, then distributed group activity would offer little advantage. Indeed, it would not really happen at all.

One powerful factor that encourages distributed group activity can be observed among members or supporters of sports teams, among students who attend the same university, and among the different racial and national groupings in the world. All foster a strong sense of group membership, where one's individual identity becomes somewhat merged with that of the group.

Besides team spirit, we are all familiar with a second factor that encourages distributed group activity from our own family life: members of the group care about each other.

Team spirit depends upon a shared experience or common interest. These may be enhanced by gossip to some extent, but probably not enough to provide a strong selection factor. Caring, on the other hand, is greatly enhanced if each member of the group knows about the lives of the others. And this is a task for which gossip is ideally suited—indeed, that is its sole purpose. The more we know about the people we encounter, the more likely we are to act on their behalf. (This is why trained terrorists who take hostages go to great lengths to avoid regular contact with their captives. They know that if they learn even a relatively small amount about their hostages, they will be unable to kill them should the situation reach that point. For the same reason, counter-terrorist agents drag out negotiations to increase the opportunity for the captors to get to know about their captives.)

Here then is a definite evolutionary advantage of gossip: it provides a mechanism for creating and maintaining group commitment. It is surely no accident that the woman who, in her lifetime, was the most gossiped about person ever—Princess Diana—was the one whom the greatest number of people around the world genuinely cared for. Millions felt her tragic death in 1997 as acutely as if she had been a member of their own family.

Given the important role of distributed group activity in human life, it seems likely that gossip formed the principal use of language (as opposed to protolanguage) from the moment it first appeared on the scene. Once again, a feature that evolved for one purpose is taken over for another. Having arisen as a side effect of off-line thinking, language was immediately hijacked to facilitate gossip.

It was certainly well suited for the task! Consider the universal structure common to all the world's languages—the universal grammar described in Chapter 6. This structure shows that all human languages are constructed, in the first place, to describe *who* did *what* to *whom* (SUBJECT-VERB-OBJECT), and secondarily to explain *how, why, where,* and *when* (the various MODIFIERS). Such a structure, along with the capacity to construct complex phrases and subordinate clauses and to combine several sentences to give a complex whole, is *exactly* what is required for gossip—to exchange the tittle-tattle of social interaction between members of the species.[1]

Frankly, I don't care what the language and communication experts say. When I see that *by far* the greatest use to which we put language is to gossip about each other, I am inclined to conclude that that is where the major evolutionary advantage of language resides.

Why? Because evolution is not about making certain behaviors *possible,* it's about making them easy and automatic. What a few individuals may be trained to do doesn't matter. Only what an entire species does, easily, naturally, and by inclination, is evolutionarily significant. That a single elephant can be trained to push a baby carriage has no meaning for

1 Indeed, syntax is so well suited for gossip that at least one expert has been tempted to suggest that language first arose precisely for that purpose. That is the view put forward by animal behaviorist Robin Dunbar in his 1997 book *Grooming, Gossip, and the Evolution of Language.* Dunbar likens gossip to the extensive grooming rituals observed in monkeys. I think Dunbar's theory has too many gaps to work as an explanation for the initial birth of language, but I find his comparison of gossip with animal grooming appealing, and somewhat supportive of my own theory of sustaining an *esprit de corps.*

evolution. That over half of all human use of language is to talk about other people is very significant indeed.

Thus, if we want to show that mathematical ability comes from our language facility, we should start out looking at what language is mostly used for. Not the way it *can* be used, with effort. Not the way it *could* be used. Not the way someone thinks it *should* be used. But what people *actually do* use it for most of the time. And that use is gossip.

OH, THAT PI—HE'S SO IRRATIONAL, YOU KNOW!

Our predilection for gossip reveals a deeply ingrained interest in the fine details of other people's lives, particularly their relationships with others. Moreover, our brains seem particularly well adapted to pursue this interest. Think for a moment of all the people you know, both in real life and in fiction, and how much you know about them—their names, their backgrounds, their interests, their hopes, their regrets, their sorrows, and their relationships to each other. That's a vast amount of information, much of it complex and interrelated. Yet apart perhaps from some of the names, you did not have to make any effort to commit all of that information to memory. What's more, the same can be said for practically everyone else, regardless of where in the world they live, their level of education, their socio-economic background, or the type of job they do.

Not only do we acquire and maintain this vast amount of information about others, we can all reason about their lives. We can have opinions on the actions of others, we can understand, explain, and pass judgment on things they do, we can guess or predict what they will do next. Again, not only can we do all of these things, we *do* do them, and without effort.

Now look back at the picture of mathematics I portrayed in Chapters 4 and 5. Mathematics studies the properties of, and relationships between, various objects, either real objects in the world (more accurately, idealized versions of those real objects) or else abstract entities that the mathematician creates.

I think we now have our answer as to what key capacity of the human brain enables some of us to do mathematics, and how (and why) our brain acquired that capacity long before mathematics came onto the scene. We have, in short, identified "the math gene." We have discovered the secret that enables mathematicians to be able to do mathematics: a mathematician is someone for whom mathematics is a soap opera.

I should stress that I am not referring to the mathematical community but to mathematics itself. The "characters" in the mathematical soap opera are not people but mathematical objects—numbers, geometric figures, groups, topological spaces, and so forth. The facts and relationships that are the focus of attention are not births and deaths, marriages, love affairs, and business relationships, but mathematical facts and relationships about mathematical objects. Are objects A and B equal? What is the relationship between objects X and Y? Do all objects of type X have property P? How many objects of type Z are there? These are the kinds of questions that interest the avid devotee of the soap opera we call mathematics.

You think that my suggestion trivializes mathematics? Not at all. As we have observed, gossip is one of nature's great evolutionary survival devices, and television soap operas can be justifiably described as "off-line gossip."

Remember too Wim Klein, the calculating wizard we met on page 69. "Numbers are friends to me," he remarked. Then, giving 3,844 as an example, he said, "For you it's just a three and an eight and a four and a four. But I say, 'Hi, 62 squared!'" Most probably you took his statements as mere whimsy. After all, numbers can't really be a person's friends, can they? Surely, no one would think of greeting a number.

But just imagine for a moment that Klein was being completely serious and absolutely literal. Imagine that, for him, numbers are indeed like people he has known a long time and has grown to love and care about. Imagine that Klein devours information about numbers the way millions of people followed the daily activities of Princess Diana or keep abreast of *Melrose Place*. Imagine that Klein, being asked the square root of 39,601, is like a Diana watcher being asked what kind of car she drove before she married Prince Charles. (The answers are 199 and a Mini Metro.)

To the ordinary person, the number π is just that: a number—what you get when you divide the circumference of a circle by its diameter. (It works out to approximately 3.14159.) But to me, π has a definite personality. It is a major character in a drama that has been unfolding on a vast landscape for the past 2,500 years. I see π as a highly influential, well-connected individual, cropping up in different locations and having relationships with many other mathematical objects. For example, π appears in computation of the most efficient way to pack spheres into a large crate, in the calculation of the odds in certain games of chance, and again when you compute the answer to various infinite sums (such as $1 - \frac{1}{3} + \frac{1}{5} - \frac{1}{7} + \frac{1}{9} - \frac{1}{11} + \ldots$, which has the answer $\frac{\pi}{4}$). Whenever I see a reference to π, I at once have a mental image of it, along with all its connections, just as mention of a hero in a novel will bring to mind a rich picture of that individual, complete with all his relationships.

The main difference between my mental image of π and my mental image of a character in a novel is that I do not have a *visual* image of π, whereas I do for a character in a novel. As a character in a mathematical drama, π is completely abstract. Its identity is its connections to other mathematical objects.

Now you know Klein's secret, and mine. It's the same secret possessed by all those people you come across who seem to be "good at math." It's not that they have different brains. It's just that they have found a way to use a standard-issue brain in a slightly different way.

To put it simply, mathematicians think about mathematical objects and the mathematical relationships between them using the same mental faculties that the majority of people use to think about other people. (In some cases, including geometry, the human brain's ability to reason about the environment is also brought into play. The overall mechanism is the same: a mental capacity developed to handle things in the real world is applied to an abstract world that the mind creates.)

Of course, observing that human mathematical ability has its origin in off-line thinking and the ability to reason about other humans and about the environment does not make these activities the same. For one thing, mathematical thinking is much narrower, more precise, and more highly focused than almost any other kind of reasoning process. We'll return to this point a little later.

Another difference is that the "characters" that inhabit the world of mathematics (i.e., abstract mathematical objects) are to be found not in the real world or on a television screen, but in the human mind, alongside all the other products of our imagination. Mathematicians work where demons lurk—deep in the interior of the mind.

But that's exactly where fictional drama is played out when we read a novel. So is there is a link between doing mathematics and reading a novel? Very possibly. In fact, when I described the way I think about the number π a moment ago, I explicitly compared mathematics to fictional drama. Similar comparisons have been made by other mathematicians, among them the late Gian-Carlo Rota of the Massachusetts Institute of Technology, who wrote the following in a 1988 article called "The Lost Café" (Cooper, *From Cardinals to Chaos,* p. 26):

> Of all escapes from reality, mathematics is the most successful ever. It is a fantasy that becomes all the more addictive because it works back to improve the same reality we are trying to evade. All other escapes—love, drugs, hobbies, whatever—are ephemeral by comparison. The mathematician's feeling of triumph, as he forces the world to obey the laws his imagination has freely created, feeds on its own success. The world is permanently changed by the workings of his mind, and the certainty that his creations will endure renews his confidence as no other pursuit.

Fiction is, of course, just one form of escapism into an artificial world. Exploring a park or a garden is another, and many mathematicians have likened their subject to the exploration of a beautiful garden. David Hilbert, probably the leading mathematician at the end of the nineteenth century, wrote of his mathematical collaboration with Felix Klein, one of the pioneers of group theory (quoted in Weyl, "David Hilbert and His Mathematical Work," p. 614):

> Our science, which we loved above everything, had brought us together. It appeared to us as a flowering garden. In this garden there are beaten paths where one may look around at leisure and enjoy oneself without effort, especially at the side of a congenial companion. But we also liked to seek out hidden trails and discovered many a novel view, beautiful to behold, so we thought, and when we pointed them out to one another our joy was perfect.

One drawback with both the fiction and the garden metaphors is that the people who write novels and design gardens have considerable freedom in which to exercise their creativity. In contrast, mathematics is highly constrained, with mathematical creativity being that of choosing what to investigate and how to carry out the investigation. I prefer to compare doing mathematics to the exploration of a grand and majestic landscape, with towering, snow-capped mountains and deep valleys filled with dense foliage. To reach the mountain peaks, you must learn all you can about the terrain in advance, formulate a good plan, assemble the right equipment, find the right path, and struggle hard to overcome all the obstacles you find in your way. Many people will get lost, others will give up and turn back. But for those who manage to reach the summit, the view is breathtaking and the sense of achievement second to none.

Non-mathematicians spend all their time in the valleys. Because of the dense forest, not only do they never get a chance to see the view from a mountain top, but they hardly ever *see* a mountain at all, other than the occasional tantalizing glimpse of a white-capped peak through the trees and the mist. For them, progress is always hard, it is easy to get lost, and they can never get any general sense of the overall terrain. Climb to the top of one mountain, however, and the view is much clearer. From there, you can see where the forest is most dense, where the path is a little easier, and how to circumvent impassable rivers. Climb two or more mountains, and you learn even more about the topography of the valleys as you look down from different angles, gaining a new perspective each time.

The mountain metaphor might help explain a point I think many nonmathematicians never realize. Advanced mathematics is generally much easier than elementary mathematics! Progress in the dense forest is always hard. But the higher you climb, the fewer the obstacles in your path, the farther you can see, and the easier it is to make headway. The only difficulty is adjusting to the rarefied air.

Moving from one peak to another (i.e., learning new mathematics) may still require a descent into the valleys, of course—even experienced mathematicians find themselves lost in the trees from time to time. But having seen the general layout of the land from one mountain top, they know roughly the direction they must follow. What is more, because they have experienced what it is like to reach one summit, and because they have looked across from that summit to all the beautiful and imposing mountains that surround it, they know that the journey will be worthwhile.

Although beautiful gardens and majestic landscapes are perhaps the most common metaphors, I earlier compared mathematics to soap operas to emphasize a point that is generally obscured: that the ability to do mathematics is a basic and fundamental one shared by all humans. Just as a great painter is a person who "simply" turns into art the same physical abilities you and I use to write our names, so too a mathematician "simply" turns into art the same mental abilities that we all use every day.

What is simple in concept may, of course, be difficult in practice. You *do* have the ability to do mathematics, but I am not claiming you will find it easy. I've been making my living doing it for over thirty years, and I still find it hard.

My use of the word "gossip" was also deliberate. Although it is literally accurate to describe the social activity I am referring to, "gossip" has very negative overtones. As I have pointed out, those overtones are in some ways completely unjustified—gossip developed as a crucial mechanism for maintaining group activity in a species whose survival depended on reliable cooperation. Leaving that issue to one side, however, the mental abilities required for gossip—*even the most socially denigrated variety*—are highly sophisticated, and already structurally adequate to support mathematical thinking.

Here is how Sir Michael Atiyah, one of the leading mathematicians in the twentieth century, describes how he has managed to be so successful as a mathematician (Minio, "An Interview with Michael Atiyah"):

> I will be thinking about something and suddenly it will dawn on me that this is related to something else I heard about last week, last month, talking to somebody. Much of my work has come that way. I go around shopping, talking to people, I get their ideas, half understood, pigeonholed in the back of my mind. I have this vast card-index of bits of mathematics from all those areas.

Do Atiyah's words remind you of anything?

Let me repeat once again, mathematics is not easy—certainly not the kind done by Atiyah. To do anything well takes effort; to do it supremely well, as Atiyah does, takes unusual talent and enormous dedication and hard work. Rather, my present goal is to identify the *origin* of the faculty that experts such as Atiyah use to such great effect. By way of comparison, few of us could run a mile in less than four minutes. Nevertheless, we know what it is like to have legs and to use them to move around. And most of us know what it feels like to run, even if only from our youth and then only slowly. We can, in short, recognize the faculties world-class athletes use in order to achieve their record-breaking performances. The difference between a top-class athlete and the rest of us is a matter of degree, not of kind.

So it is with mathematics. Mathematicians are not born with an ability that no one else possesses. Practically everybody has "the math gene," just as practically everybody is born with two legs. We have it because the features of our brains that enable us to do mathematics are the same ones that allow us to make sense of the world and the people in it: the large repertoire of types that we use to classify the world and the syntactic structure the human brain acquired when it became able to think off-line.

BARBIE WAS RIGHT

A few years ago, Mattel, Inc., makers of the children's doll Barbie, angered the mathematical education establishment when they brought out a new range of Barbies that spoke a number of phrases, among them "Math is hard." The furor was so strong that Mattel quickly eliminated the phrase from the doll's repertoire. The objectors were right to complain—discouraging girls from attempting mathematics aggravated an educational problem that was already bad. But many objectors went a step further, claiming that "math is not hard, just different." I think they had it completely backward. Barbie was right: math *is* hard. Anyone who tries to claim otherwise doesn't understand what mathematics is. But it's not different at all.

Leaving aside (for now) many people's self-fulfilling expectations, two things make mathematics hard. They both arise because, when we do mathematics, we are taking features of our brains that evolved for one purpose and using them for another.

The first difficulty is that we are taking an ability—indeed, an instinct—to see and reason about patterns and relationships in the physical and social world, and applying it to an abstract world of our own creation. As

Wason's test (page 115–17) demonstrates, we find it much more difficult to reason about an abstract situation than about a logically analogous real situation. Evolution has given us a brain that can think off-line, but it was essentially a one-step development. We can reason off-line about real objects or things that resemble real objects—what I have called level 3 abstraction—because our brains can generate activation patterns sufficiently like those that result from sensory input.

To do mathematics, however, we need to think off-line about completely abstract objects that bear virtually no relationship to anything in the real world—level 4 abstraction. We need to generate brain activation patterns unlike anything that arises from sensory input. Now, we know that practically everyone can do this, since that is precisely what is required to have a sense of number and to cope with various abstractions that arise in modern life, such as the concepts of marriage, ownership, or indebtedness. Still, we generally find it hard. From a subjective point of view, the brain becomes comfortable with a new abstraction by making that abstraction seem more concrete. It does this through the familiarity that comes from repeated exposure. And that takes hard work.[2]

The second feature of mathematics that makes it hard is the degree of rigor required in its reasoning processes. As I argued in Chapters 2 and 3, precise reasoning is not something for which our brains evolved. But we need to be careful in drawing conclusions from this observation. Precise, formal reasoning is not required for mathematical *discovery*. Rather, its purposes are verification of things already discovered (or perhaps suspected) and convincing others of the truth of those discoveries.

2 This observation should give pause for caution to those educational reformists who assert that there is little or no value in repetitive, routine, "drill" exercises. What little we know about the way the brain works, in particular how it acquires new skills, suggests the very opposite. This is not to say that mathematics education should consist solely, or even largely, of routine, repetitive work. But drill is probably not a dispensable component of mathematical education. Repetitive practice is as important in learning how to do mathematics as it is in learning how to play tennis or to play the guitar.

The need for formal verification is a direct consequence of the nature of mathematical discovery. Trial and error, guesswork, intuition, and conversations with others can go on for days, months, even years, with key steps often being carried out while the mathematician is either asleep or thinking about something else. Although this process need not be mere haphazard stumbling—for a good mathematician it can be highly focused and efficient—it can nevertheless generate errors. Formal proofs are the final (and totally reliable) safeguard against false "discoveries."

In Chapter 5, I likened solving a mathematical problem to finding the best arrangement of furniture in a new house. To do that, we generally proceed by a combination of experience, intuition, guesses, and trial and error, rather than step-by-step logical deduction. This point was articulated in 1968 by the world-famous mathematician Paul Halmos ("Mathematics as a Creative Art"):

> Mathematics—this may surprise you or shock you some—is never deductive in its creation. The mathematician at work makes vague guesses, visualizes broad generalizations, and jumps to unwarranted conclusions. He arranges and rearranges his ideas, and he becomes convinced of their truth long before he can write down a logical proof.

Only after the mathematician thinks she has solved a problem does she start to work out a logical proof, a process that involves organizing the various ideas and insights that led to the solution into a precise, logical argument. This part of the process requires a great deal of training.

Richard Borcherds, who won a Fields Medal, the mathematician's equivalent of the Nobel Prize, in 1998, made a similar point. In an interview just after being awarded the prize, Borcherds declared (Borcherds and Gibbs, "Monstrous Moonshine Is True"):

> The logical progression comes only right at the end, and it is in fact quite tiresome to check that all the details really work. Before that, you have

to fit everything together by a lot of experimentation, guesswork, and intuition.

Again, mathematicians insist on precise, logical proofs because the discovery process can lead to errors. Because they work in a completely abstract realm, mathematicians cannot check their results by observation, as astronomers do, or by performing experiments, like chemists. Logical proof is the arbiter of truth in mathematics, and that is where the logical thinking is required.

But for all its difficulty, doing mathematics does not require any special ability not possessed by every one of us. What distinguishes a great mathematician from a high school student struggling in a geometry class is the degree to which the mathematician can cope with abstraction, the complexity of the abstract world that the mathematician can create and hold in his mind, the level of insight he can attain about that world, and his skill in carrying out the reasoning processes and formulating a logical proof.

WHY DO SO MANY PEOPLE SAY THEY CAN'T DO MATH?

Why then do so many people say they are simply *unable* to do math? There is no single answer. In mathematics, as in everything else, people vary. The point I have been making in this book is that everyone has the basic ability—we all have "the math gene."

Most people could get much further than they think they can. Mathematics is not unlike marathon running. Before the 1970s, only a few highly trained athletes ran marathon races. The ability to run non-stop for 26.2 miles was thought to require special talent. Most people find it difficult to run one mile, and before 1970 or thereabouts, fit sports men and women seldom ran much beyond five miles at a time. Then the running boom took off, first in the United States and then in many other countries. Before long, thousands of ordinary people around the world were running marathons. They took far longer to finish than the elite world-class runners. But they were finishing.

Running a marathon did not require a special talent after all. For most people, all it takes to run a marathon is a sufficient desire to do so. Talent only matters if you want to do it better than everybody else.

Similarly for mathematics. The key to being able to do mathematics is wanting to. I am not talking about becoming a great mathematician or venturing into the heady heights of advanced mathematics. I am speaking solely about being able to cope with the mathematics found in most high school mathematics curricula (using a calculator to compensate for the brain's unsuitability for doing arithmetic).

Two pieces of evidence support my claim. First, when people find that they really need to master some mathematics, they invariably do so.

Here's a concrete example. At the start of the twentieth century, the French psychologist Alfred Binet examined two calculating wizards who made their living demonstrating their arithmetical prowess. As we saw in Chapter 3, the main secret behind the number wizards' amazing performances is that numbers have meanings for them. Binet wanted to see how these two professional calculators performed when compared with another group for whom numbers had meaning—cashiers in a store. (This was in the days before mechanical cash registers were common, let alone electronic check-out machines.)

Binet arranged a calculating contest between the two professional calculating wizards and four cashiers at the Paris department store Bon Marché. The result? The cashiers easily outperformed the prodigies at basic arithmetic. For example, to compute the product 638 × 823, it took the better prodigy 6.4 seconds, whereas the best cashier got the answer in just 4 seconds; when asked to calculate 7,286 × 5,397, the better prodigy took 21 seconds against just 13 seconds for the best cashier.

Let me give you one more example. (Again, it is an arithmetical example, not the higher math that is my main focus here, but that only makes the point all the more dramatic.) In the early 1990s, a group of psychologists

(Nunes, Schliemann, and Carraher, *Street Mathematics and School Mathematics*) tested the arithmetical abilities of a group of Brazilian third-grade schoolchildren who, when out of school, worked in a street market. All the children had been taught the standard procedures for addition, subtraction, multiplication, and division.

When the researchers gave them a standard school arithmetic test, the students did poorly, averaging as low as 14 percent on simple subtraction problems involving whole numbers of three or fewer digits. Yet, when they were working at their market stalls, the children did just fine.

For example, to compute the change from 200 cruzeiros for a purchase costing 35 cruzeiros, one child (as tape-recorded by a researcher posing as a customer) proceeded like this:

> If it were thirty, then the result would be seventy. But it's thirty-five. So it's sixty-five. One hundred sixty-five.

Let's look at what this particular subject was doing. First he splits the 200 into 100 + 100. (He doesn't vocalize this step, but it's clear that this is what he did.) He puts one 100 to one side and sets out to compute 100 – 35. To do this, he first rounds off 35 to 30, and computes 100 – 30. This he can do easily: the answer is 70. Then he corrects for the rounding by subtracting the 5 he ignored: 70 – 5 = 65. Finally, he adds the 100 he had put to one side at the beginning: 65 + 100 = 165. Not only does he get the right answer quickly—in his head in a noisy street market—but he employs some nifty mathematical manipulations to do so. In fact, a mathematician would say that the boy's solution employs more sophisticated mathematical thinking than simply applying the standard school subtraction algorithm.

Was this boy the one junior Einstein in the street market? No, he was entirely typical. Although they were hopeless on math tests at school, all the children were equally fluent in the calculations they needed at their market stalls. As the researchers put it, the children were hopeless at school mathematics but superb at street mathematics. What's the difference between the two? Not the mathematics itself: 2 + 2 = 4 on the street just as much as in the classroom. The difference was that when they were working at their market stalls, they had a strong motivation to do the calculations, and the numbers meant something to them.

So much for needing a special brain to be able to do mathematics!

My second piece of evidence that the key to mathematics is wanting to is simply the nature of mathematical thought. Leaving aside the problems associated with learning the multiplication table and getting the right answer in arithmetic—at which mathematicians are no better than anyone else—mathematics involves much the same off-line thought processes used in everyday life. The difference is that, in mathematics, the off-line thought is focused on objects that are themselves pure abstractions, whereas in everyday life our thoughts generally focus on real objects or fictional versions of real objects—level 4 abstraction as against level 2 or level 3. Mathematics thus gives the mind the additional burden of having to create an entire world and hold it in memory. Everyone can do this—the mental facility required is precisely the facility for off-line symbolic thought that gives us language. But people do seem to vary in how far they can take that facility—a difference in the extent and complexity of the abstract world that we can create and hold in our minds.

A typical soap opera creates an abstract world, in which a highly complex web of relationships connects fictional characters that exist first only in the minds of the program's creators and are then recreated in the minds of the viewer. If you were to think about how much human psychology, law, and even everyday physics the viewer must know in order to follow and speculate about the plot, you would discover it is considerable—at least as much as the knowledge required to follow and speculate about a piece of modern mathematics, and in most cases, much more. Yet viewers follow soap operas with ease. How are they able to cope with such abstraction? Because, of course, the abstraction is built on an extremely familiar framework. The characters in a soap opera and the relationships between them are very much like the real people and relationships we experience every day. The abstraction of a soap opera is only a step removed from the real world. The mental "training" required to follow a soap opera is provided by our everyday lives.

In contrast, it requires considerable conscious effort to train the mind to follow the soap opera we call mathematics. The characters in the mathematical soap opera (i.e., the various entities that mathematicians study) do not resemble things in our everyday lives. Even though the relationships between those objects are generally very similar to familiar relationships in the everyday world, those relationships *seem* strange and unfamiliar. Mathematics becomes possible for the mathematician because she spends sufficient time in the abstract world of mathematics for it to achieve a degree of reality for her. But whereas the real world constantly reinforces the abstract world of the television soap opera, the mathematician herself must provide that reinforcement for the mathematical soap opera.

If you ask the very best mathematicians how they go about their work, they will talk in terms of "exploring" or "experiencing" a "world." The better mathematicians they are, the more real that world seems to them. This is why they can "know" that a certain result is true long before they are able to construct a logical proof. What attracts them to mathematics is a fascination with that particular world. In that respect, mathematics is no different than biology, playwriting, acting, sports, or any other pursuit.

What exactly gives rise to a fascination with mathematics in particular is a question for which I don't have a good answer. I can, however, offer suggestions as to why some people may be attracted to mathematics.

One aspect of mathematics that attracts many is that it is neatly ordered and reliable, very much in contrast with the everyday world we all live in. An individual who seeks a high degree of order in his life might well find it in mathematics. (On the other hand, a person who seeks to exercise *control* over an ordered world is likely to be frustrated by mathematics, which rules its universe according to its own laws.)

Culture and peer pressure also influence people's perceptions of mathematics and, in consequence, their ability to do it. The most obvious illustration of this is the high proportion of Jewish men who become successful mathematicians, far higher relative to their numbers than any other ethnic-gender group. This is surely a consequence of growing up in home, cultural, and often neighborhood environments that put a high value on learning, and on mathematics in particular. Hungary provides another example of a society that puts a high value on the learning of mathematics, and which, as a result, enjoys a higher level of mathematical ability than most other nations. At the other end of the spectrum, the poor performance in higher mathematics exhibited by teenage girls and young women in the United States until recently was generally acknowledged to be a result of peer pressure and society's expectations of their ability. (These days, women are starting to outnumber men among mathematics majors at American colleges and universities.)

Whatever it is that causes the interest, it is *that interest* in mathematics that constitutes the main difference between those who can do mathematics and those who claim to find it impossible.

As I suggested earlier, doing mathematics is very much like running a marathon: it does not require any special talent and "finishing" is largely a matter of wanting to succeed. But why should anyone want to enter the mathematical race in the first place?

RIGHT ANSWER, WRONG REASON

Why do all modern societies insist that children learn mathematics? The most common justification is that the world is so dependent on science and technology that everybody needs to be good at mathematics in order to lead a successful life and contribute fully to society. That's like saying that because our lives are so dependent on the automobile, everyone should be able to fix a car. While an automobile-dependent society certainly requires an adequate number of well-trained auto engineers and mechanics, for most of us it's enough to know how to drive. Similarly for mathematics.

The "you-need-it-to-get-by-in-a-technological-society" argument has a distinguished history. In America, for instance, it was put forward by Benjamin Franklin, who founded the Philadelphia Academy of Science in 1751. But it was relatively recently that a mechanism was established to provide general mathematics (and science) education for all American children. It began in the early years of this century, largely as a result of the educational theories of John Dewey, who spoke of developing in all students "scientific habits of the mind." Speaking at a 1909 symposium, Dewey said ("Symposium on the Purpose and Organization of Physics Teaching in Secondary Schools, Part 13"):

> Contemporary civilization rests so largely upon applied science that no one can really understand it who does not grasp something of the scientific methods and results that underlie it; . . . a consideration of scientific resources and achievements from the standpoint of their application to the control of industry, transportation, communication...increases the future social efficiency of those instructed.

Although Dewey's goal of teaching "scientific habits of the mind" was never achieved on any appreciable scale, he and his followers had a major impact on the school curriculum, where mathematics and science have ever since ranked right after English as the most important subjects.

It is, of course, obvious why everybody should take English: in today's society, everyone needs to be literate and able to communicate well. But when everybody can afford a pocket calculator and a great many successful people have little knowledge of mathematics or science, the traditional justification for mandatory mathematics and science education simply does not stand up. I'll give you two arguments that I think do.

The first does not lend much support to traditional mathematics and science education. In fact, it suggests that we should design courses that demonstrate what mathematics and science are and what role they play in modern life, rather than teach particular skills. These courses would more closely resemble typical history or social studies courses than existing mathematics and science courses. Their arithmetical and manipulative-algebraic content and "problem-solving" work would be included to give a feel for what is involved, and the goal would be completion of the task, not a "perfect performance" or "getting the right answer."

Courses of this kind are, I suggest, entirely consistent with the main goal of education, which is not to train people for a particular job or career but to pass on thousands of years of human culture and learning from one generation to the next. This purpose appears to have been forgotten in today's credential-laden society, where the key to a better life is often not so much education as the possession of an appropriate diploma. But the main purpose of education was clearly recognized in the past.

For instance, in *The Republic,* Plato wrote that every citizen should have an appreciation of mathematics and an understanding of the nature of the world. It should be added, however, that a citizen in ancient Greece was part of the privileged elite. Vocational training, in the form of apprenticeships, was the most anyone else could hope for.

A similar appreciation that education was intended to pass on human culture (and was not vocational training) also prevailed in Europe until well into the twentieth century. As in ancient Greece, such an education was reserved for a relatively small elite. Moreover, in the European case, science was classified as vocational, not part of what was required to be accepted as an "educated person."

The importance of education as opposed to vocational training follows from the way the human brain evolved. Our survival trick for hundreds of thousands of years has been our rich repertoire of categories (or types) for carving up the world, coupled with our instinct and our ability to acquire knowledge, to question things we see, understand how things work, design and manufacture tools, formulate plans, collaborate with one another, and communicate in a symbolic (not merely indexical) fashion.

These abilities and instincts are all passed on from generation to generation through our genes. But the knowledge and skills required to make use of those abilities and instincts are not.[3]

On the other hand, the benefits of passing on knowledge and skills are manifestly obvious: it makes no sense for each new generation to keep reinventing the wheel. Since the knowledge and skills that we acquire are not passed on genetically, we have developed another mechanism for doing so: education. The ability to pass on the accumulated knowledge of each generation to the next is what has led to the rapid advancement of human society. This is why human life and society have changed so dramatically in the past 5,000 years, even though the brain has hardly changed at all.

Of course, human beings—and probably only human beings—are not slaves to their genes, so we have to be careful about using evolutionary arguments to support lifestyle choices or public policy decisions. When the human race acquired off-line thinking, it acquired the ability to make deliberate choices.

Since the principal key to our success as a species (and our success as individuals in that species) is our ability to learn and to adapt to changing circumstances, it is surely best to devote the initial, and greater, part of any education to *developing that ability to learn and adapt to changing circumstances.* That argues for a broad education of the kind generally referred to as "liberal arts." Vocational training, which is by definition narrow and function-specific, can come later.

Today, because of the rapid acceleration of societal change, existing skills can become obsolescent and new skills may be required every few years. Practically any vocational training that would have been beneficial

3 The idea that abilities and skills learned during the lifetime of one individual are incorporated into that person's genes and then passed on to the next generation biologically was widely believed in the last century, and is generally referred to as Lamarkianism after one of its original proponents. It was long ago demonstrated to be false.

to the parents of today's teenagers would be of no use to their children, who can expect to have to acquire new skills roughly every seven years throughout their working lives. Accordingly, the best survival technique that we can provide for our children is the ability to acquire new knowledge and learn new skills.

Part of that armory of survival skills is a general understanding of mathematics and an ability to acquire specific mathematical skills as and when required. That means equipping people to take a mental faculty that they already possess and use in their everyday social lives, and teaching them how to use that faculty in the abstract realm of mathematical thinking.

A second, more mundane reason to learn mathematics can be found in Dewey's phrase "scientific habits of the mind." For the vast majority of people, the subject matter of mathematics and science is not as important as the mode of thinking. In science, that means the need to collect and weigh evidence, base decisions on that evidence, think logically, and be willing to change one's opinion on the basis of new evidence. In math, it means—well, by now you know pretty well what constitutes mathematical thinking. Why is it good for you?

In 1997, the U.S. Department of Education released a white paper (the Riley Report, named after the Secretary of Education) highlighting the importance of pre-college mathematics for gaining entrance to college and success on the job market, especially for low-income students. Using data from several long-term studies, that report found that 83 percent of high school students who took algebra and geometry courses went on to college, more than double the rate (36 percent) of students who did not take those courses. Low-income students who took algebra and geometry were almost three times as likely to attend college as those who did not. Moreover, students who had completed those courses did noticeably better at college than their colleagues who did not.

The report did not say anything about getting a good grade in those mathematics courses, or even about passing. Simply taking the courses led to the benefits. What's more, students got the same benefits regardless of the subjects that they pursued at college. English, history, and art students benefited along with the mathematics and science majors. It's the thought process that makes the difference.

The message conveyed to us by the Riley Report is clear: a daily dose of mathematical thinking is as good for the mind as a daily walk or jog is for the body.

Unlike my first argument, this second argument for mathematics and science education implies practice in actually *doing* mathematics and science. Educational traditionalists might be tempted to plump for my second argument and ignore the first. To do so would be to miss a great opportunity to dramatically improve the performance in mathematics and science of a large section of our population, and (as we have just seen) along with it their overall intellectual ability. Here is why.

Most of us are not motivated—at least, not for long—by arguments that something is good for us. If we were, we would all eat far less and exercise more. Learning mathematics and science is hard and involves some tedious, repetitious activity. Of course, the same can be said of many other pursuits—learning to play games and sports, learning to play a guitar, learning to skateboard, or learning to act. And yet people learn to do these things willingly. Why? What persuades them to keep at it during the difficult, initial learning stage when progress seems elusive and frustration is high? The answer is obvious. They keep going if they can see what benefit lies at the end. Similarly for mathematics and science. If our schools would devote some time to showing their pupils the fascination of these subjects, many more would be motivated to put in the effort to master the basic techniques.

Throughout the above discussion I have concentrated on the mathematical education of the general population. We do much better with mathematically talented individuals who want to pursue the subject in depth. Contrary to what you sometimes read in the press about a chronic shortage of mathematicians, there has for many years now been a surplus of qualified mathematicians, giving the mathematical profession a far higher unemployment rate than many other fields. So long as we provide a constantly open door to advanced mathematical education, society will not face a shortage of mathematical expertise. The shortage is in people who have an adequate *general* ability in mathematics alongside various other skills. And the key to addressing that very real and critical shortage is first to arouse their interest and second to make it possible for them to acquire sufficient mathematical skill by developing educational programs that take into account the way the brain does mathematics.

And there I shall rest my case, and step down from the pulpit.

· 10 ·

ROADS NOT TAKEN

WHAT IS THE STATUS of the thesis that I have advanced in this book? Does our ability to do mathematics really depend on the same features of the brain that give us language? And did the faculty for language arise through a single mutation in the hominid brain sometime within the last 200,000 years?

I have presented an argument that I hope you find reasonably convincing. But no one knows the answer to either question for sure. Since minds do not leave any fossil evidence, we may never know. Even if a perfectly preserved body of one of our *Homo erectus* ancestors were one day found frozen in a glacier, with its brain intact, it would probably not give us the answers.

This is the great scientific frustration: the very thing that appears to be the major distinguishing feature of human beings—our great evolutionary survival trick—seems to be the one feature whose origin we may never know.

The problem is not, let me stress, because there is any uncertainty about the process of evolution by natural selection. That theory is as well established as any scientific theory we have, and indeed better than most. Not only is evolution itself well established as the framework for the development of life on earth, but we know with confidence a great many details about the evolution of many species, including humans.

However, although the framework of evolution is beyond doubt, many arguments about *particular* evolutionary developments remain speculative. Not because there is any serious doubt that life evolved, but because we cannot perform experiments to test particular evolutionary hypotheses. With evolution, there was only one "experiment," namely, what actually happened. Fortunately, that "experiment" provided us with an enormous amount of evidence, only a fraction of which has so far been discovered, so we do have one way to test our theories: we can use those theories to make predictions as to what kinds of evidence we might be able to find, and then look for that evidence.

In the case of *Homo sapiens,* whereas we have reasonable confidence in our understanding of the lineage of our species, that confidence does not extend to the acquisition of language or mathematical ability. Any explanation for this process has to be consistent with both the historical and fossil evidence and with our knowledge of the abilities of present-day humans and other species. We should also apply Ockham's razor, and make our explanations as simple as possible, based on the minimum number of (plausible) assumptions. Provided that we proceed in such a fashion (and I maintain that in this book I did just that), anyone who thinks we have it wrong will have an obligation to come up with a better explanation —a better fit with the known facts and with Ockham's razor.

I am not aware of any other serious attempt to explain the origins of human mathematical ability at the level of detail provided in this book. There have been a number of attempts to describe the evolutionary development of numerical ability or "number sense," but as I explained earlier, that is not at all the same as mathematical ability. On that basis, my explanation that mathematical ability is essentially just a new use of a mental faculty that gave us language—namely, off-line, symbolic thinking—is, by default, "the best current explanation." Certainly, it fits all the known facts. Moreover, I claim that it is highly plausible. In addition, I really can't think of any other explanation that is both plausible and fits the facts. But for all that, the field is wide open.

The suggestion that language comes automatically with the ability for off-line, symbolic thought, on which my explanation for mathematical ability rests, is due originally to Bickerton. It assumes that off-line thinking involves syntactic structure as a "model of the world." Maybe there is another reasonable notion of symbolic thinking that does not require syntax, but no one has produced one so far, and I can't imagine what it would look like.

Things get more murky when we come to the origins of off-line thought itself. Is the human brain—which is beyond doubt qualitatively different from other brains in its functionality—the result of a sudden mutation—possibly the Linguistic Eve scenario—or could it have arisen through a process of gradual change across a whole population? There are several explanations that seem as though they might hold water, some of which I described in Chapter 8. At present, the evolution of language is an area of heated debate in linguistics.

I should re-emphasize that my argument about mathematical ability being a novel use of our linguistic faculty does not in any way depend on the origins of our linguistic ability. Thus, even if you were to take away the three arguments I gave—Linguistic Eve, small change, big effect, and the attractor theory—the rest of my argument would not be at all weakened. (Some details of the overall plot would change, of course.)

Another part of my argument where it is possible to take an alternative path is the reason for the enormous growth in the hominid brain that preceded the acquisition of language. Recall that, although the brain's acquisition of syntax seems to have occurred rapidly—perhaps even suddenly—the development that preceded it took place over the 3,500,000-year period during which the hominid brain tripled in size. I said I believed that the principal driving force behind this growth was the need for increased cognitive capacity—what I referred to as types and the ability to categorize, and which Bickerton equivalently called protolanguage.

Of course, that does not have to have been the whole story. Other factors may also have contributed to brain growth. One factor that I mentioned was the acquisition of color vision. But could color vision have been the *primary* driving force behind brain growth, as some have suggested? I suspect not, and here's why. Since apes also have color vision, it's likely that our most recent common ancestors had it as well. But that was long before the hominid period. Color vision can only account for a much earlier period of brain growth in the primate line—although it might explain in large part why primates have larger brains than other species.

Another possible factor driving brain growth was suggested by the neurophysiologist William Calvin in the early 1990s: throwing (see Calvin, "The Unitary Hypothesis"). I find Calvin's idea plausible, at least as a major contributing factor. It is possible that both throwing and protolanguage together drove the growth of the brain and prepared the way for language. On the other hand, Calvin goes further than claiming that accurate throwing contributed to hominid brain growth. He suggests it was the *primary* cause, and that accurate throwing set the stage for the acquisition of language.

Here, briefly, is Calvin's argument.

A good way of felling an animal in order to capture it and kill it for food is to throw a rock at it. One way to do this would be for the hunter to hide in the bushes or behind the rocks at a waterhole and wait for suitable prey to arrive. Instead of rushing out with spears, which would cause the prey to scatter, the hunter would down it with a hurled stone.

The success of such a strategy clearly depends on being able to throw accurately. Pinpoint accuracy is not necessary, Calvin suggests, because many species of animals go to waterholes in groups. If a hunting hominid could hit one of a group, there was a good chance that at the very least it could be knocked off balance. The animals might well stampede, knocking down and trampling some of the herd, thus doing part of the hunter's work. (With a group of hunters all throwing at once, the likelihood of a stampede is much greater.)

Accurate throwing requires a large brain on at least two counts, Calvin maintains. First, the speed of the throwing action is too great for the muscles involved to be controlled by the sensory feedback process used for slower actions such as picking up an egg or raising a cup to your lips. (It takes about a tenth of a second for a signal to travel from an arm sensor to the spinal cord and back out to the arm muscles, which is not much less than the total length of time it takes to throw a rock.) The instructions for throwing have to be prepared in advance and then executed in a precise sequence.

Calvin's second argument that a large brain is necessary for throwing is that it requires a significant amount of computation to produce a throw that can hit a target the size of, say, a dog at a distance of thirty paces. Since greater accuracy in throwing would clearly confer survival advantage, selection would favor those hominids with a better "waterhole kill average" and in this way would drive the development of the larger brains required to produce more accurate throws.

Calvin's throwing accuracy thesis does not conflict with the theory of increasing cognitive capacity and sophistication that I described earlier, and it may even be a contributing factor. But I have trouble accepting the throwing thesis as the major factor driving brain growth, because I see no reason why accurate throwing should require the enormous brain possessed by hominids. Throwing is a single skill that can be acquired with repetitive training, and very small brains are able to perform one or two specific activities well. For example, a frog can fling out its tongue and catch a passing fly half a body-length away, a rattlesnake can spring forward and catch a mouse a couple of feet away, and a cat can spring from a low tree-branch to catch a bird several feet away. Where a large brain is required is to perform a wide variety of functions and, in particular, to produce a "dog that can keep learning new tricks."

Calvin links his throwing idea to the more general notion of (in his words) "stringing things together." After observing that the ability to string things together is predominantly lodged in the left brain, he gives the following examples ("The Unitary Hypothesis"; emphasis in the original):

- For toolmaking and tool use, one usually needs to make a novel *sequence* of movements.
- To go beyond the categories of ape language...and say something novel, our ancestors seem to have hit upon the trick of assigning meaning to a sequence of individually meaningless vocalizations.
- To create a novel plan of action, one needs to spin a scenario connecting the past with the future. And then contemplate it before acting . . .
- This "get set" phase is particularly important for versatile throwing; one has to produce trains of nerve impulses going to dozens of muscles, timed like the finale of a fireworks display sequence.
- ...music...Instead of stringing phonemes together to make words and sentences, we string chords together to make phrases and melodies.

For Calvin, "stringing things together" is the key mental capacity behind language and the various other uniquely human faculties.

The problem with this proposal is that the ability to string things together is a necessary condition for language and thinking off-line, but not a sufficient one. As I explained in Chapter 6, grammatical structure is about nesting, not stringing things together in a line, and similarly for modeling the world in off-line thought.

Of course, Calvin is aware of this objection, and he has an answer. In its favor, Calvin's answer is a Darwinian one. Suppose that the brain generates many different sequences, and only the "fittest" (or "best-adapted" or "most appropriate") one wins out and gains acceptance. In that case, argues Calvin, the recognizable nested structures that constitute the rules of grammar are simply one way of describing the linear sequences of words that survive the unconscious Darwinian selection process in the brain.

That might indeed be a reasonable description of the mechanism that the human brain uses to generate and comprehend (or more precisely, parse) grammatical utterances. If so—if indeed the brain generates many, if not all, possible sequences of words and then selects the ones that work (i.e., that give a grammatical sentence)—then the key question is: how is the selection done? I would say that a brain that is able to make that selection is a brain that has syntax built in. And now we are back at the original question: how did the hominid brain first acquire that syntactic structure? Calvin's proposal provides a possible implementation of the process that the brain actually follows, but it does not answer that fundamental question.

Of course, I did not provide a "definitive" explanation either. Rather, I recounted several plausible hypotheses, each with some supporting evidence.

One of those hypotheses was Linguistic Eve. While I personally see no reason to reject this hypothesis, to many people it seems "obviously false." Let me try to counter what I think is the reasoning—or, more accurately, the intuition—that leads them to this conclusion.

The heart of the issue, I think, is that the suggestion that all humans alive today are descendants of a single human only seems surprising from a present-day standpoint, when the world population numbers billions. Let me elaborate.

Today's huge population is a consequence of our having hit—very recently—the steep rise of an exponential growth curve. If you were to start to generate a gigantic family tree of all humans alive today, you would not have to go back many generations before the numbers start to shrink dramatically. By the time you get back to the middle part of the *Homo sapiens* period, you would find a few thousand or so individuals living at any one time—10,000 is the figure often given in the anthropological literature. (Remember, many individuals do not live to child-bearing age, and many others leave a descendant line that eventually peters out.) Trace those few thousand back toward the start of the *Homo sapiens* line and you will surely come to a single individual, a last common ancestor. (For the Linguistic Eve hypothesis, it's essential that there be a common *Homo sapiens* ancestor, of course. If you have to go even further back to find a common ancestor, there could not be a single Linguistic Eve.)

The above argument is essentially a mathematical one, not biological. It's a question of the structure of tree hierarchies, of which genealogical or family trees are good examples. If you follow such a tree downward (i.e., into the future), which is the way people usually think of them in an evolutionary context, repeated branching gives you an exponentially increasing population. Looking even a few generations into the future, the possibilities seem endless. But when you start to move backward in such a tree structure (i.e., into the past), following, say, the female line, a very different pattern emerges. First, large sections of the tree never appear, namely all those individuals whose female-descendant line petered out before the present day. Second, as you go back, the tree shrinks steadily. Since no one can have two biological mothers, the total number of individuals that you encounter at each generation can never increase. (Remember, we have already pruned out individuals not having a present-day descendant along the female line.) On the other hand, the tree can get "narrower," which is what will happen whenever a female has two or more female offspring whose female-descendant lines extend to the present day. And whenever the tree becomes narrower, it can never grow any broader. Eventually, therefore, you will find yourself with a tree made up of a very small number of individuals.

This mathematical argument alone does not guarantee that the tree will come down to a single individual. But it does, I think, make the idea of a Linguistic Eve numerically acceptable.

Another factor that contributes to our surprise at the Linguistic Eve proposal is our terribly poor ability to judge the likelihood of various eventualities involving coincidences. A classic example is the birthday problem, which asks how many people you need to have at a party so that there is a better-than-even chance that two of them will share the same birthday. Most people think the answer is 183, the smallest whole number larger than 365/2. In fact, you need just 23. (The answer 183 is the correct answer to a very different question: how many different birthdays have to be represented at a party so that there is a better-than-even chance that one of them will be *your* birthday?) The figure of 23 arises because there is no restriction on which two people will share a birthday. This makes an enormous difference. Similarly, with the human ancestral tree, there is no restriction on how far back you need to go to find a common ancestor, just that there is *some* stage at which the tree converges to a single individual.

Finally, what of my choice of richer representation as the main driving force behind the emergence of language, as opposed to richer communication?

First, notice that I did not say that one came before the other. It seems highly likely that the two developed in tandem. Rather, my point is that in order to properly understand how we acquired language, we should view it as a representational structure rather than as a medium of communication. In order to communicate some concept, you first need to have a mental representation of it. (Okay, so it works the other way if someone tells you something. But when you trace that idea back to its source, it had to have started out inside someone's head, not in their vocal tract!)

Moreover, although communication was important from the moment our ancestors started to cooperate, throughout the long period of protolanguage growth, increased understanding (i.e., richer representations) was surely a greater evolutionary pressure.

Still another reason for viewing language as a representational medium is that this is what our description of off-line thinking says it is! Readers familiar with the theories of the great Russian psychologist L. S. Vygotsky will no doubt recognize similarities between off-line thought and Vygotsky's suggestion that all purposeful abstract thought is an internalization (or mental objectification) of physical action. Vygotsky and his followers use this hypothesis as a basis for their educational theories, suggesting in particular that the only way to achieve mastery of abstract concepts and processes is through repeated drill—a possibility I put forward in Chapter 9.

Vygotsky's views on the nature of purposeful thought bear on the issue of whether cognition or communication was the primary driving force behind the acquisition of language. For Vygotsky, the only distinction between thinking and using language is vocalization. Syntax is, therefore, essential for purposeful thought. When a person is engaged in conscious thought, he is carrying on an internal conversation with himself. Thus, if we accept Vygotsky's teaching, the question as to whether the evolutionary pressure for language acquisition was thought or communication disappears.

EPILOGUE
How to Sell Soap

The studio boss looked at the fresh and eager faces seated around the long conference table at Plato Television Studios. Jabbing his long, fat cigar toward them to punch home each point, he began:

"You guys are the best series scriptwriters in town. I've invited you here today to offer you the chance of a lifetime: to work on the best television series the world has ever seen. Here's the outline of what I have in mind."

He glanced at the young woman sitting at the far end of the table by the side projector. "Alice, can you give us the first slide, please?"

Alice dimmed the lights and switched on the projector. The studio boss continued in the half-light:

"There are these two families, you see: the Points and the Lines. Basically, the series is about what these two families do, and the relationship between them. One of the great things about this idea is that I've set it up so that the series is going to be really cheap to make. As you can see on the slide, we don't need to hire any actors for the Points, because the Points have no parts."

The studio boss used his cigar to gesture toward the words on the screen at the first sentence, which read:

Points have no parts.

"The idea," he continued, "is to use digital special effects to represent the Points. Those computer graphics guys are cheap these days—every time the aerospace industry downsizes or a computer company is bought out by Microsoft, another thousand of them are thrown out of work and we can pick them up for a song."

There was an audible sigh of relief around the room as the eleven scriptwriters realized how easily they could have been the ones forced to change careers in mid-life, but for the fortunate accident that they had flunked Algebra 2 in high school and studied English Lit instead.

"We'll also save a ton of money on the Lines," the studio boss beamed, obviously pleased at his ingenuity. "My idea is that a Line has no breadth. That means we don't need trained actors. We can use more of those out-of-work schmucks from aerospace." All eyes turned to the second item written on the slide, as the studio boss jabbed at it with his cigar. It read:

A Line has no breadth.

"Now, you guys are the best, so I'm going to give you a lot of freedom on this project," the boss growled. "All you have to do is stick to five guiding principles for the way I want the series to go. Next slide, Alice."

Everyone looked as the next slide appeared on the screen.

"I'll give you each a copy of this," said the studio boss, "but let me summarize the five items you can see.

"Item one. You can have a Line that connects any two Points. That should be clear enough.

"Number two. You can continue any Line as long as you want. No problem there, either. Hell, they do that in lots of long-running series.

"The next one might need a bit of explaining. As you can see, what it says is 'Any Point can be the center of a Circle of any size.' My idea is that each episode can center around a particular person in the Point family. That episode will concentrate on the circle of friends of that person. Some weeks, it might be a small circle, other weeks a really big one—and since the circle will be made up of Points, it doesn't make any difference to the budget."

There was a chorus of appreciation around the table as the scriptwriters began to see the potential of the studio boss's overall idea to keep the budget down.

"Next one: 'All right angles are the same.' Every series has to have an angle. You all learned that in Television 101. You also know that some angles are right for the intended audience, some are wrong. In my series, there's just one right angle. All other angles are wrong, and anyone who tries to introduce one will be off the show faster than I can say 'soap.' Understood?"

The studio boss looked around, daring anyone to respond to his challenge. No one did. The thought of all those unemployed former aerospace workers was still fresh in their minds, and they did not want to join them in the line for unemployment benefits.

"I'm not entirely sure I need the fifth guideline," said the studio boss, who was clearly enjoying showing off the genius of his new idea. "It might be superfluous, given the other four. But I put it down just to be sure. It's a bit hard to follow—I'll get Alice to work on the text. But what it boils down to in simple terms is this: if one of you sets it up so that two of the Lines are not supposed to meet, then no matter who else takes over the storyline in a later episode, those two Lines are still not going to meet. Ever! *Capisce?*"

Everyone nodded. The boss leaned back in his chair and chewed on his cigar.

"That's it. Any questions?"

There was silence for a moment, then a young woman halfway along the table raised her hand. "I think it's a great concept," she began, "but I've got one question."

"Fire away," replied the studio boss.

"How long do you expect this to run? Thirteen weeks? Fifty-two? Or are you planning on something that goes on for years, like *As the World Turns* or *General Hospital?*"

The studio boss chuckled. "Little lady, I don't think you've quite got the message as far as my overall concept is concerned." He gestured toward the five guiding principles on the screen. "This idea is so insanely great, it has so much potential, that it's going to change the world. Believe me, once it catches on and we get enough sponsors, this baby is going to run for thousands of years. Or my name's not Euclid."

APPENDIX

The Hidden Structure of Everyday Language

In Chapter 6, I said that every grammatical sentence of any human language can be constructed by the repeated application of a single rule for "putting things together," coupled with rules for the order in which certain things go (e.g., subject before verb before object) and rules for agreement (e.g., singular/plural, gender, and tense). Whereas the rules for order and agreement vary from language to language, the basic "putting things together" rule is the same for all languages and is best expressed by means of the tree shown in Figure A.1.

I have been calling this structure the fundamental language tree. Linguists call it the "X-bar tree," and the theory of syntactic structure that shows how this tree can produce all grammatical sentences is called "X-bar theory." In this appendix, I will illustrate the underlying ideas of X-bar theory (as well as explain the origin of the theory's rather peculiar name). In particular, I will show how each of the different kinds of phrase

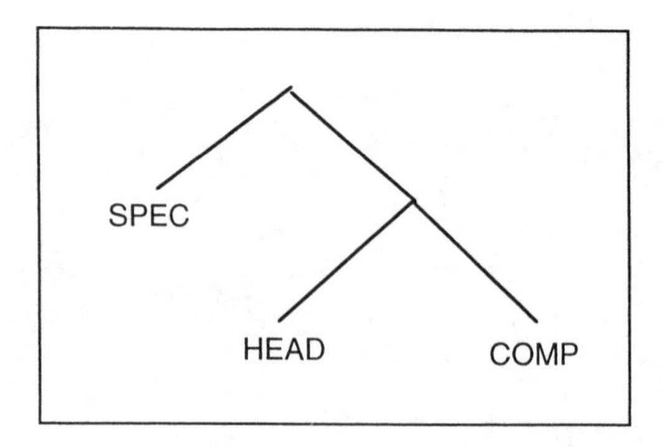

FIGURE A.1 The Fundamental Language Tree: The Generic Structure for X-bar Theory

can be built up using just the X-bar—noun phrases, verb phrases, adjectival phrases, prepositional phrases, even sentences themselves. Although they may appear to fit together in different ways, in reality they all have the same basic structure. That structure involves three parts.

First, every phrase has what is called a *head*. The head is the only constituent of a phrase that has to be present; the two other constituents are optional. The head cannot be longer than a single word. (It can be less, as we shall see.) The head of a noun phrase must be a noun; the head of a verb phrase must be a verb or a part of a verb; the head of an adjectival phrase must be an adjective. The head exercises considerable control over the clause.

The head of a phrase is linked first to its *complement*, if it has one. For example, in the noun phrase *the house with the red door*, the head is the noun *house* and its complement is the prepositional phrase *with the red door*. Together, the noun (N) and the prepositional phrase (PP) that is its complement (N-Comp) form what is known as the N-bar, *house with the red door*. The terminology "N-bar" stems from the notation generally used to denote this construction: an N with a bar over it, as shown in Figure A.2. To complete the noun phrase (NP), the N-bar is then combined with the *specifier* (Spec-N) *the*. This construction is shown in Figure A.2. (The triangle beneath the node PP in Figure A.2 is a standard way to

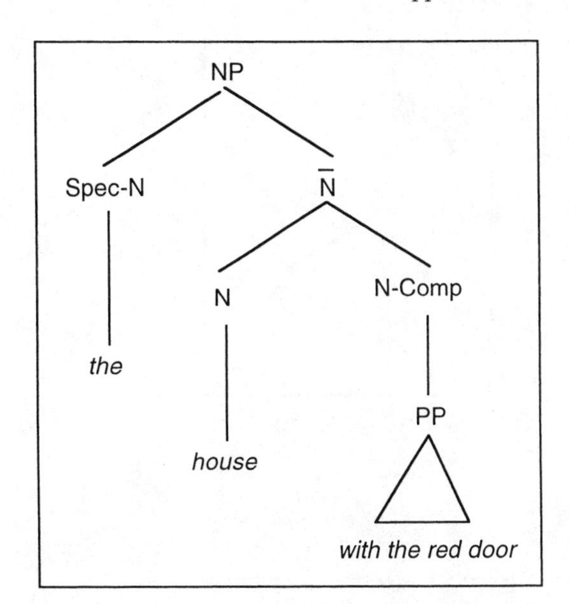

FIGURE A.2 Construction of a Noun Phrase

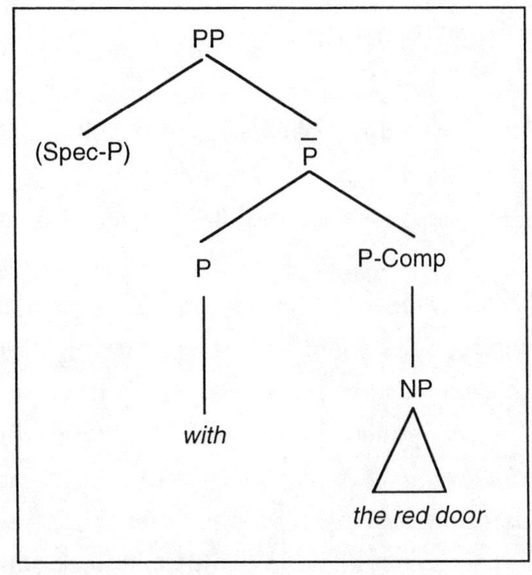

FIGURE A.3 Construction of a Prepositional Phrase

indicate that this is itself a phrase that requires further analysis. The triangle is intended to suggest a further parse tree.)

Likewise, the prepositional phrase *with the red door* has its own phrase structure tree, illustrated in Figure A.3. The head of this phrase is the preposition *with*. Notice that for this phrase, there is no specifier. This is

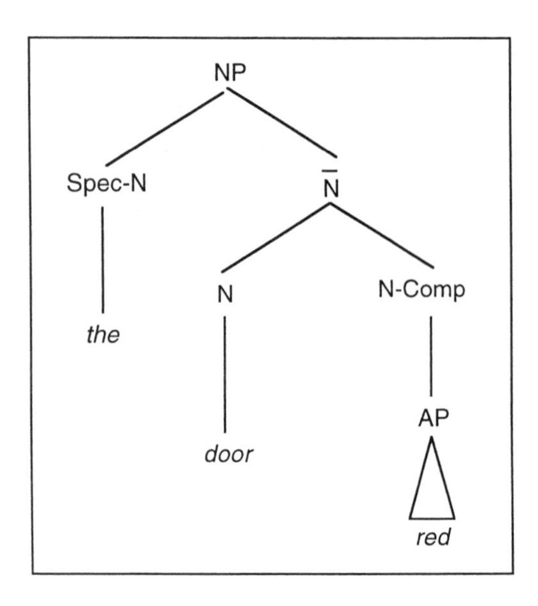

FIGURE A.4 Construction of the Noun Phrase *the red door*

indicated by putting parentheses around the label Spec-P (as well as by there being no entry beneath the label).

Figure A.4 illustrates the next step in the analysis, the structure of the noun phrase *the red door.*

Notice that, since any complement has to be a phrase, the N-Comp *red* is considered an adjectival phrase, not a word. This means that another step is required to complete the analysis: the phrase structure for the AP *red,* given in Figure A.5. (This is why there is a triangle beneath the AP node in Figure A.4.)

Notice that by consistently drawing my trees with the specifier to the left, then the head, then the complement, I have drawn the phrase structure tree for *the red door* so that the horizontal ordering of the tree (which, you will recall, is not a structural feature of the tree) differs from the ordering of the words in English. (It does, however, agree with the ordering in French: *la porte rouge.*)

I should perhaps stress that I was not being overly pedantic in insisting that the role played by the complement *red* in the noun-phrase *the red*

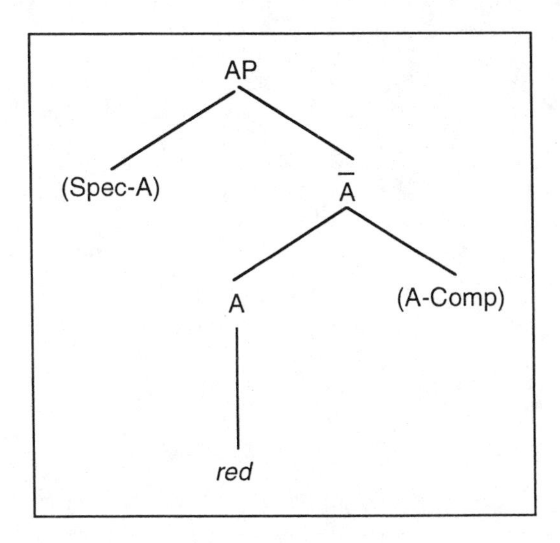

FIGURE A.5 Construction of the Adjectival Phrase *red*

door is that of an adjectival *phrase,* not a word. When describing a mechanism for putting linguistic units into boxes to give larger linguistic units (which may themselves be put into still-larger boxes), we are looking for a simple mechanism that works in all cases. The approach that linguists have taken is to regard any phrase as made up of a head—which is no longer than a single word—an optional complementary *phrase,* and an optional specifier. This means that we occasionally have to put a single word into a box.

Actually, we allow a head to be combined with more than one complementary phrase, as occurs with the noun phrase *the house with the red door where Sally lives.* Figure A.6 gives the parse tree for that phrase. Each complement is itself a phrase (i.e., a box).

Figure A.7 shows the general combinatory structure for constructing phrases. In that diagram, we use the letter X as a sort of variable, to stand for any N (noun), V (verb), P (preposition), A (adjectival/adverbial), etc. The parentheses indicate an optional item. The asterisk indicates an item that can be repeated (in principle, any finite number of times; in prac-

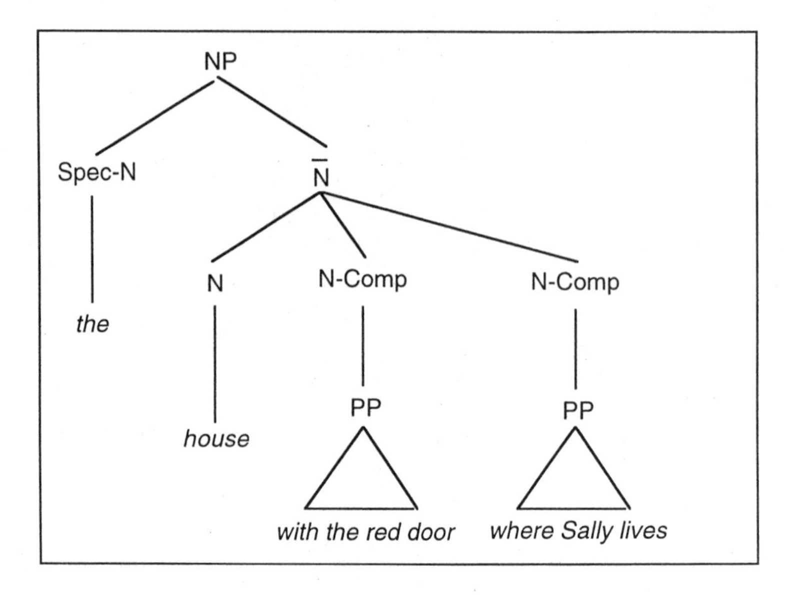

FIGURE A.6 Structure of a Noun Phrase with Two Complements

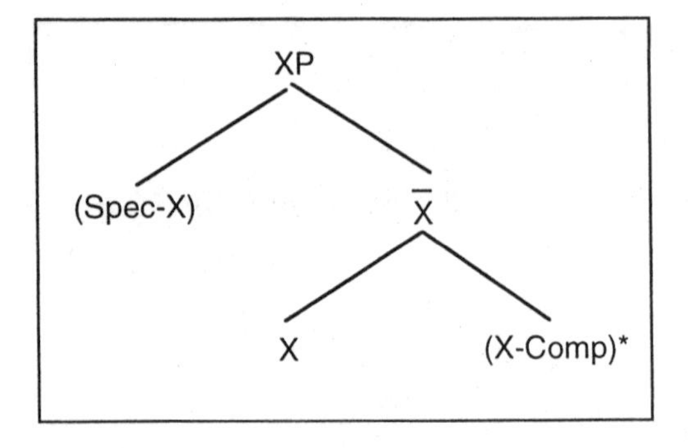

FIGURE A.7 The Generic Structure for X-bar Theory

tice, generally just once or twice). Because the key step is the combination of the head X and the complement phrase (or phrases) X-Comp to give the unit whose label is read aloud as "X-bar," the theory of phrase structure that I have just outlined is known among linguists as "X-bar theory."

The phrase structure trees of X-bar theory clearly indicate the possible interpretations of an ambiguous phrase. For example, the two possible readings of the phrase *the house with the red door that Max likes* are illustrated in Figure A.8. In the first reading (top), Max likes the house; in the second reading (bottom), he likes the door. In both examples, the analysis is incomplete; each of the complement phrases on the bottom line requires further "unpacking."

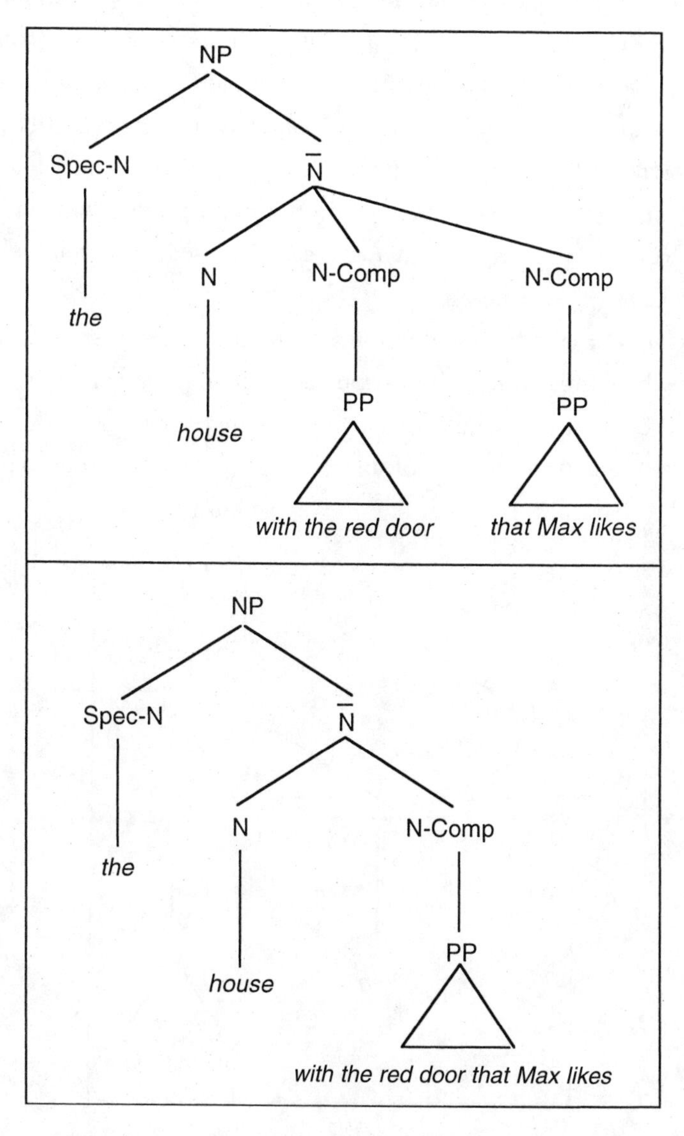

FIGURE A.8 Disambiguation of an Ambiguous Noun Phrase

Figure A.9 gives an example of a verb phrase, *run into the room*. Again, we see the familiar X-bar combination of the head of the phrase (in this case, an untensed verb), a verb complement (in this case, a prepositional phrase), and no specifier.

Other kinds of phrase are handled similarly. But what about an entire sentence or clause? Isn't that also a phrase (or possibly a logical combination of phrases, connected together by *and, or, if . . . then, unless,* etc.)?

Let's start with a simple example: *John ran into the room*. If this is to fit our pattern of phrase structure, it must have a head. But what is that head? It can't be *John*. *John* is a noun, and only noun phrases have nouns as heads. Nor can it be *ran*, since *ran* is a (cased, tensed) verb, and only verb phrases have verbs as heads.

The answer is to be found in the very nature of a sentence. Sentences are the whole units where the various parts come together to give informational content—a claim about the world in the case of a declarative sentence, a request in the case of an interrogative, etc. In order to provide

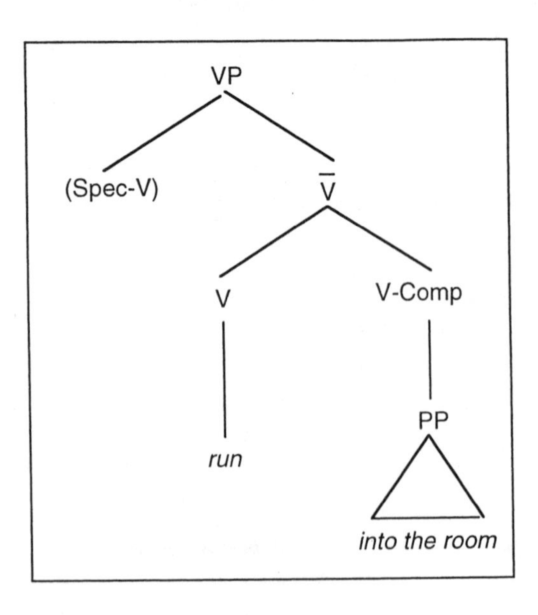

FIGURE A.9 The X-bar Structure of a Verb Phrase

informational content, a sentence must specify the tense of the main verb. And that tense is the head. Or rather, tense is part of the head. The verb must also have a case, which must agree with the subject of the sentence, and that too is part of the head of the sentence. The tree for our example *John ran into the room* is given in Figure A.10.

Since the tense and case of many verbs is provided by an *inflection* to the verb, it is common to use the letter I (or sometimes the abbreviation INFL) for the head, as I have done. The subject of the sentence, necessarily a noun phrase, is then the specifier for I, Spec-I (in our example, the NP *John*). The I-Comp is the untensed, uncased verb phrase (*run into the room* in our example). The tensed, cased verb phrase (in our example, *ran into the room*) is then the I-bar combination. The entire sentence, S, could also be denoted as IP, the *inflection phrase.*

It is particularly important to recognize that the verb phrase *run into the room* is not tensed or cased. The tense and case of the main verb in a sentence is provided by the head. This view of a sentence puts considerable

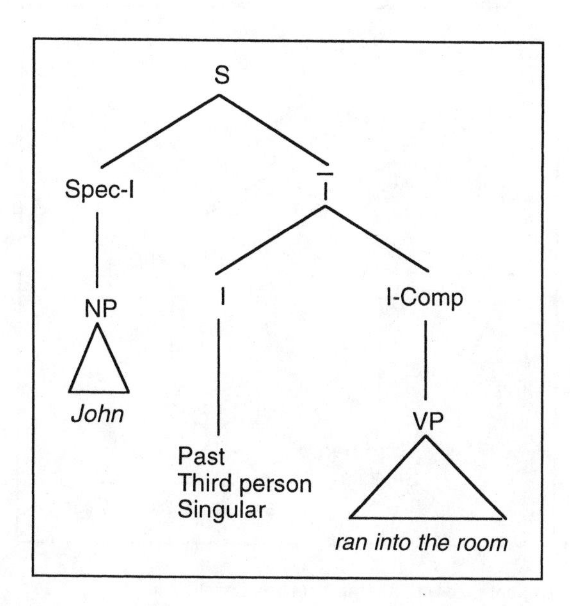

FIGURE A.10 A Parse Tree for a Sentence

emphasis on the tense and case of the main verb. As I have mentioned, the head of any phrase has a governing role over the remainder of the phrase. But this is exactly what happens with the tense and case of the main verb in a sentence. Tense reaches into the main verb phrase (the I-complement), generally causing an inflection to the main verb, while case reaches across to the subject (the I-specifier), ensuring agreement between the subject and the main verb.

This example illustrates how sentences may be constructed using the same combinatorial machinery that gives all other phrases. This is a desirable feature both for scientific parsimony and for helping us understand how *Homo sapiens* could have acquired syntax in the first place. If it seems odd to think of a sentence in this way, part of the reason may be that this is the first example we have encountered where the head is not a word. I could have avoided this by taking a slightly different example: *John did run into the room.* Now the head is a clearly identifiable word, *did.* Of course,

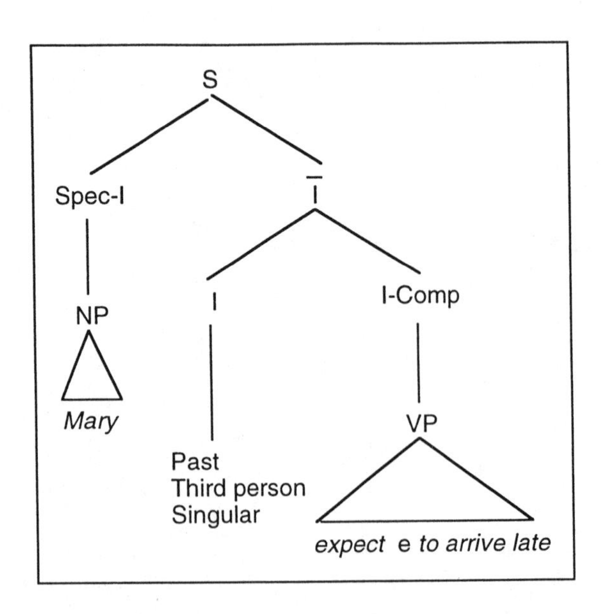

FIGURE A.11 A Sentence with an Unarticulated Constituent

did still carries tense and case, which govern the rest of the sentence just as before. In this case, INFL has a word of its own.

In fact, there are other examples where an important constituent of a phrase is not explicit. For example, the sentence *Mary expected to arrive late* has the phrase structure illustrated in Figure A.11.

The person that Mary expected to arrive late was obviously Mary herself. The sentence has an unarticulated constituent in the position denoted by the letter *e* in the following, linguist's version of the original sentence: *Mary expected* e *to arrive late.* As shown in Figure A.11, the unarticulated constituent *e,* which denotes Mary, plays a role in the grammatical structure of the sentence. The constituent *e* continues to play a role as we further analyze the verb phrase *expect* e *to arrive late.*

REFERENCES

CHAPTER 1: A MIND FOR MATHEMATICS

I have written extensively for a general audience on the material in this chapter. The following four books all provide background material and further reading on the topics covered. They are listed in increasing order of difficulty.

Devlin, Keith. 1998. *Life by the Numbers*. New York: John Wiley. The official companion to the six-part PBS television series of the same name, for which I was an adviser.

———. 1994. *Mathematics: The Science of Patterns: The Search for Order in Life, Mind, and the Universe*. Scientific American Library Series. New York: W. H. Freeman.

———. 1998. *The Language of Mathematics: Making the Invisible Visible*. New York: W. H. Freeman.

———. 1999. *Mathematics: The New Golden Age*. 2nd ed. New York: Columbia University Press.

CHAPTER 2: IN THE BEGINNING IS NUMBER

Much of the experimental work described in this chapter—and a great deal more as well—can be found in the book *The Number Sense: How the Mind Creates Mathematics* by Stanislas Dehaene (New York: Oxford University Press, 1997). The specific source of the material in Dehaene's book on multiplication tables is on p. 127.

Another excellent source is *The Mathematical Brain* by Brian Butterworth (Basingstoke, UK: Macmillan, 1999). The specific sources of material in Butterworth's book are: Charles, pp. 267–71, 301–8; Signora Gaddi, pp. 163–67; Frau Huber, pp. 170–72; Julia, pp. 300–301.

A standard reference for work on children's mathematical development is the book *The Child's Understanding of Number* by R. Gelman and C. R. Gallistel (Cambridge, MA: Harvard University Press, 1978).

Tobias Dantzig's book *Number: The Language of Science*, first published in 1954, was republished in 1967 (New York: Free Press).

All of the specific experiments referred to in the chapter were reported in the articles and books listed below.

Antell, S. E., and D. P. Keating. 1983. "Perception of Numerical Invariance in Neonates." *Child Development* 54, pp. 695–701.

Bijeljac-Babic, R., J. Bertonici, and J. Mehler. 1991. "How Do Four-Day-Old Infants Categorize Multisyllabic Utterances?" *Developmental Psychology* 29, pp. 711–21.

Boyson, S. T., and G. G. Bernston. 1996. "Quantity-Based Interference and Symbolic Representations in Chimpanzees (*Pan trogolodytes*)." *Journal of Experimental Psychology: Animal Behavior Processes* 22, pp. 76–86.

Brannon, E. M., and H. S. Terrace. 1998. "Ordering of the Numerosities 1 to 9 by Monkeys." *Science* 282, pp. 746–49.

Church, R. M., and W. H. Meck. 1984. "The Numerical Attribute of Stimuli." In H. L. Roitblat, T. G. Bever, and H. S. Terrace, eds., *Animal Cognition* (Hillsdale, NJ: Erlbaum).

Koechlin, E., S. Dehaene, and J. Mehler. 1997. "Numerical Transformations in Five-Month-Old Human Infants." *Mathematical Cognition.*

Koehler, O. 1951. "The Ability of Birds to Count." *Bulletin of Animal Behavior* 9, pp. 41–45.

Kronecker, L. *Jahresberichte der Deutscher Mathematiker Vereinigung, Bd.* 2, p. 19.

Matsuzawa, T. 1985. "Use of Numbers by a Chimpanzee." *Nature* 315, pp. 57–59.

McComb, K., C. Packer, and A. Pusey. 1994. "Roaring and Numerical Assessment in Contests Between Groups of Female Lions, *Panthera leo*." *Animal Behavior* 47, pp. 379–87.

McGarrigle, J., and M. Donaldson. 1974. "Conservation Accidents." *Cognition* 3, pp. 341–50.

Mechner, F. 1958. "Probability Relations Within Response Sequences Under Ratio Reinforcement." *Journal of the Experimental Analysis of Behavior* 1, pp. 109–21.

Mechner, F., and L. Guevrekian. 1962. "Effects of Deprivation upon Counting and Timing in Rats." *Journal of the Experimental Analysis of Behavior* 5, pp. 463–66.

Meck, W. H., and R. M. Church. 1983. "A Mode Control Model of Counting and Timing Processes." *Journal of the Experimental Psychology: Animal Behavior Processes* 9, pp. 320–34.

Mehler, J., and T. G. Bever. 1967. "Cognitive Capacity of Very Young Children." *Science* 158, pp. 141–42.

Pepperberg, I. M. 1987. "Evidence for Conceptual Quantitative Abilities in the African Gray Parrot: Labeling of Cardinal Sets." *Ethology* 75, pp. 37–61.

Piaget, J. 1952. *The Child's Conception of Number.* New York: Norton.

———. 1954. *The Construction of Reality in the Child.* New York: Basic Books.

Simon, T. J., S. J. Hespos, and P. Rochat. 1995. "Do Infants Understand Simple Arithmetic?: A Replication of Wynn (1992)." *Cognitive Development* 10, pp. 254–69.

Starkey, P., and R. G. Cooper, Jr. 1980. "Perception of Numbers by Human Infants." *Science* 210, pp. 1033–35.

Starkey, P., E. S. Spelke, and R. Gelman. 1983. "Detection of Intermodal Numerical Correspondences by Human Infants." *Science* 222, pp. 179–81.

———. 1990. "Numerical Abstraction by Human Infants." *Cognition* 36, pp. 97–127.

Woodruff, G., and D. Premack. 1981. "Primitive Mathematical Concepts in the Chimpanzee: Proportionality and Numerosity." *Nature* 293, pp. 568–70.

Wynn, K. 1992. "Addition and Subtraction by Human Infants." *Nature* 358, pp. 749–50.

———. 1995. "Origins of Numerical Knowledge." *Mathematical Cognition* 1, pp. 35–60.

———. 1996. "Infants' Individuation and Enumeration of Actions." *Psychological Science* 7, pp. 164–69.

CHAPTER 3: EVERYBODY COUNTS

John Allen Paulos's book *Innumeracy: Mathematical Illiteracy and Its Consequences* was first published by Hill & Wang (New York) in 1988.

As in the previous chapter, many of the psychological examples cited here are described in Stanislas Dehaene's book *The Number Sense* and/or Brian Butterworth's book *The Mathematical Brain,* as are many other illustrations of the points I make. The specific sources of material in Dehaene's book are: Paris patient with brain lesion, p. 69; multiplication tables, p. 127. The specific sources of material in Butterworth's book are: Donna, pp. 197–99; Dottore Foppa, pp. 196–97; Signora Gaddi, pp. 163–67; Frau Huber, pp. 170–72.

The articles describing the specific experiments mentioned are listed below.

Dehaene, S., S. Bossini, and P. Giraux. 1993. "The Mental Representation of Parity and Numerical Magnitude." *Journal of Experimental Psychology: General* 122, pp. 371–96.

Dehaene, S., E. Spelke, P. Pinel, R. Stanescu, and S. Tsivkin. 1999. "Sources of

Mathematical Thinking: Behavioral and Brain-Imaging Evidence." *Science* 284, pp. 970–74.

Gleason, A. 1984. "Mathematics: How Did It Get to Where It Is Today?" *Bulletin of the American Academy of Arts and Sciences* 38 (October), pp. 8–24.

Hauser, M. D., P. MacNeilage, and M. Ware. 1996. "Numerical Representations in Primates." *Proceedings of the National Academy of Sciences USA* 93, pp. 1514–17.

Henik, A., and J. Tzelgov. 1982. "Is Three Greater Than Five?: The Relation Between Physical and Semantic Size in Comparison Tasks." *Memory and Cognition* 10, pp. 389–95.

Miller, K., C. Smith, J. Zhu, and H. Zhang. 1995. "Preschool Origins of Cross-national Differences in Mathematical Competence: The Role of Number Naming Systems." *Psychological Science* 6, pp. 56–60.

Moyer, R. S., and T. K. Landauer. 1967. "Time Required for Judgments of Numerical Inequality." *Nature* 215, pp. 1519–20.

Schmandt-Besserat, D. 1989. "Oneness, Twoness, Threeness." *The Sciences* (New York Academy of Sciences), pp. 44–48.

Washburn, D. A., and D. M. Rumbaugh. 1991. "Ordinal Judgments of Numeral Symbols by Macaques." *Psychological Science* 2, pp. 190–93.

CHAPTER 4: WHAT IS THIS THING CALLED MATHEMATICS?

Devlin, K. 1998. *Life by the Numbers.* New York: John Wiley.

———. 1998. *The Language of Mathematics: Making the Invisible Visible.* New York: W. H. Freeman.

Mandelbrot, B. 1988. *The Fractal Geometry of Nature.* New York: W. H. Freeman.

Murray, J. 1980. "A Pattern Formation Mechanism and Its Application to Mammalian Coat Markings." In *Lecture Notes in Biomathematics*, 39 (Heidelberg, Ger.: Springer-Verlag), pp. 360–99.

———. 1988. "Mammalian Coat Patterns: How the Leopard Gets Its Spots." *Scientific American* 256, pp. 80–87.

Prusinkiewicz, P., and A. Lindenmayer. 1996. *The Algorithmic Beauty of Plants.* Heidelberg, Ger.: Springer-Verlag.

Steen, L. 1988. "The Science of Patterns." *Science* 240, pp. 611–16.

Sawyer, W. W. 1955. *Prelude to Mathematics.* London: Penguin Books.

Thompson, D'Arcy. 1961. *On Growth and Form.* Cambridge: Cambridge University Press (first published in 1917).

Turing, A. 1952. "The Chemical Basis of Morphogenesis." *Philosophical Transactions of the Royal Society,* B237.

Wigner, E. 1960. "The Unreasonable Effectiveness of Mathematics in the Natural Sciences." *Communications in Pure Applied Mathematics* 13, no. 1.

CHAPTER 5: DO MATHEMATICIANS HAVE DIFFERENT BRAINS?

Hadamard, J. 1945. *The Psychology of Invention in the Mathematical Field.* New York: Dover.

Hoffman, P. 1987. "The Man Who Loves Only Numbers." *Atlantic Monthly* (November).

Littlewood, J. E. 1978. "The Mathematicians' Art of Work." *Mathematical Intelligencer* 1, no. 2, p. 114.

Paulos, J. A. 1987. *Innumeracy: Mathematical Illiteracy and Its Consequences.* New York: Hill & Wang.

Penrose, R. 1989. *The Emperor's New Mind: Concerning Computers, Minds, and the Laws of Physics.* Oxford: Oxford University Press.

Russell, B. 1910. *The Study of Mathematics: Philosophical Essays.* London: Unwin Books.

Synge, J. L. 1957. *Kandelman's Krim.* London: Jonathan Cape.

Watson, P. C., and P. N. Johnson-Laird. 1972. *Psychology of Reasoning: Structure and Content.* Cambridge, MA: Harvard University Press.

CHAPTER 6: BORN TO SPEAK

To my mind, the best introduction by far to contemporary linguistics is Steven Pinker's book *The Language Instinct: How the Mind Creates Language* (New York: William Morrow, 1994).

Chomsky, N. 1957. *Syntactic Structures.* The Hague: Mouton.

———. 1965. *Aspects of the Theory of Syntax.* Cambridge, MA: MIT Press.

Crain, S., and M. Nakayama. 1986. "Structure Dependence in Children's Language." *Language* 62, pp. 522–43.

CHAPTER 7: THE BRAIN THAT GREW AND LEARNED TO TALK

Bickerton, D. 1990. *Language and Species.* Chicago: University of Chicago Press.

———. 1995. *Language and Human Behavior.* Seattle: University of Washington Press.

Devlin, K. 1997. *Goodbye Descartes: The End of Logic and the Search for a New Cosmology of the Mind.* New York: John Wiley.

Savage-Rumbaugh, S., S. Shankar, and T. Taylor. 1998. *Apes, Language, and the Human Mind.* New York: Oxford University Press.

Tobias, P. V. 1971. *The Brain in Hominid Evolution.* New York: Columbia University Press.

CHAPTER 8: OUT OF OUR MINDS

Barwise, J., and J. Perry. 1983. *Situations and Attitudes.* Cambridge, MA: MIT Press.

Bickerton, D. 1990. *Language and Species.* Chicago: University of Chicago Press.

———. 1995. *Language and Human Behavior.* Seattle: University of Washington Press.

———. 1998. "Catastrophic Evolution: The Case for a Single Step from Protolanguage to Full Human Language." In J. Hurford, M. Studdert-Kennedy, and C. Knight, eds., *Approaches to the Evolution of Language* (Cambridge: Cambridge University Press, 1998), pp. 341–58.

Byrne, R., and A. Whiten. 1985. "Tactical Deception of Familiar Individuals in Baboons." *Animal Behavior* 33, pp. 669–73.

———, eds. 1998. *Machiavellian Intelligence: Social Expertise and the Evolution of Intellect in Monkeys, Apes, and Humans.* Oxford: Clarendon Press.

Cann, R., M. Stoneking, and A. Wilson. 1987. "Mitochondrial DNA and Human Evolution." *Nature* 325, pp. 31–37.

Carstairs-McCarthy, A. 1998. "Synonymy Avoidance, Phonology and Origin of Syntax." In J. Hurford, M. Studdert-Kennedy, and C. Knight, eds., *Approaches to the Evolution of Language* (Cambridge: Cambridge University Press, 1998), pp. 279–96.

Chomsky, N. 1975. *Reflections on Language.* New York: Pantheon Press.

Devlin, K. 1991. *Logic and Information.* Cambridge: Cambridge University Press.

Gould, S. J., and R. Lewontin. 1979. "The Spandrels of San Marco and the Panglossian Paradigm: A Critique of the Adaptionist Program." *Proceedings of the Royal Society* B205, pp. 581–98.

Hurford, J., M. Studdert-Kennedy, and C. Knight., eds. 1998. *Approaches to the Evolution of Language.* Cambridge: Cambridge University Press.

Lieberman, P. 1992. *The Biology and Evolution of Language.* Cambridge, MA: Harvard University Press.

MacNeilage, P., M. Studdert-Kennedy, and B. Lindblom. 1984. "Functional Precursors to Language and Its Lateralization." *American Journal of Physiology* 246, pp. R912–14.

McPhail, E. 1987. "The Comparative Psychology of Intelligence." *Behavior and Brain Sciences* 10, pp. 645–95.

Peirce, C. S. 1940. 1940. *Philosophical Writings of Peirce.* Edited by J. Buchler. New York: Dover.

Premack, D., and G. Woodruff. 1978. "Does the Chimpanzee Have a Theory of Mind?" *Behavioral and Brain Sciences* 4, pp. 515–26.

Rumelhart, D., and J. McClelland. 1986. *Parallel Distributed Processing.* Cambridge, MA: MIT Press.

Sacks, O. 1990. *Seeing Voices: A Journey into the World of the Deaf.* London: HarperCollins.

Whiten, A., and R. Byrne. 1988. "The Manipulation of Attention in Primate Tactical Deception." In R. Byrne and A. Whiten, eds., *Machiavellian Intelligence: Social Expertise and the Evolution of Intellect in Monkeys, Apes, and Humans* (Oxford: Clarendon Press, 1998), p. 218.

CHAPTER 9: WHERE DEMONS LURK AND MATHEMATICIANS WORK

Borcherds, R., and W. Gibbs (author). 1998. "Monstrous Moonshine Is True." *Scientific American* (November), pp. 40–41.

Cooper, N. G., ed. 1988. *From Cardinals to Chaos.* Cambridge: Cambridge University Press.

Dewey, J. 1909. "Symposium on the Purpose and Organization of Physics Teaching in Secondary Schools, Part 13." *School Science and Mathematics* 9, pp. 291–92.

Dunbar, R. 1997. *Grooming, Gossip, and the Evolution of Language.* Cambridge, MA: Harvard University Press.

Halmos, P. 1968. "Mathematics as a Creative Art." *American Scientist* 56 (Winter), p. 380.

Minio, R. 1984. "An Interview with Michael Atiyah." *Mathematical Intelligencer* 6, p. 16.

Nunes, T., A. Schliemann, and D. Carraher. 1993. *Street Mathematics and School Mathematics.* Cambridge: Cambridge University Press.

Weyl, H. 1944. "David Hilbert and His Mathematical Work." *Bulletin of the American Mathematical Society* 50.

CHAPTER 10: ROADS NOT TAKEN

You will find a good introduction to the evolution of language in *Approaches to the Evolution of Language*, edited by James Hurford, Michael Studdert-Kennedy, and Chris Knight (Cambridge: Cambridge University Press, 1998). Other sources referred to in this chapter are:

Calvin, W. 1993. "The Unitary Hypothesis: A Common Neural Circuitry for Novel Manipulations, Language, Plan-Ahead, and Throwing?" In K. R. Gibson and T. Ingold, eds., *Tools, Language, and Cognition in Human Evolution* (Cambridge: Cambridge University Press, 1993), pp. 230–50.

Chomsky, N. 1975. *Reflections on Language.* New York: Pantheon Press.

Dawkins, R. 1976. *The Selfish Gene.* Oxford: Oxford University Press.

Gibson, K. R., and T. Ingold, eds. 1993. *Tools, Language, and Cognition in Human Evolution.* Cambridge: Cambridge University Press.

Vygotsky, I. S. 1962. *Thought and Language.* Cambridge, MA: MIT Press.

INDEX

R

W

X

A NOTE ABOUT THE TYPE

The Math Gene has been composed in Monotype Fournier, a digital type based on the letterforms of 18th-century designer-printer Pierre Simon Fournier. The youngest son of a French printing family, Fournier established his own foundry in 1742. By the end of his life he is reputed to have cut 60,000 punches for 147 of his own alphabets. Fournier's greatest achievement, however, was the development of the *transitional* style, which connected the earlier Baroque types to those of the modern period, such as those of Bodoni. Fournier's *St Augustin Ordinaire*, the type used in his 2-volume *Manuale Typographique* (1764/1768), became the model for the Monotype design, which was released in metal in 1925.

Book design and composition by Mark McGarry,
Texas Type & Book Works, Inc.